Biogas Technology in the Third World:

A Multidisciplinary Review

Andrew Barnett
Leo Pyle
S. K. Subramanian

Contents

About the Authors

Andrew Barnett received his MA in Development Economics from the University of Sussex and then specialized in the problems of cost-benefit analysis, first with the Ministry of Overseas Development, London, and then as a free-lance consultant. At the time of writing this report he was at the Institute of Development Studies, University of Sussex, Brighton. He is currently Advisor to the Science and Technology Policy Programme of the International Development Research Centre at the Science Policy Research Unit, University of Sussex.

Leo Pyle obtained his PhD from Cambridge University in Chemical Engineering. He has taught at Imperial College, London for the last 10 years during which time he was Visiting Professor both at Queen's University, Kingston and the University of Chile. He has carried out research on anaerobic fermentation kinetics and is currently Chairman of the Methane and Fertilizer Working Group of the Intermediate Technology Development Group, London.

S.K. Subramanian gained his PhD from Birmingham University in Chemical Engineering and Petroleum Technology and worked in the chemical industry in India for a number of years. Following a period as Assistant Director of the National Chemical Laboratory, Poona, he moved to the Department of Science and Technology of the Government of India and was Secretary to the Indian National Council of Science and Technology. Later, he became Director of Technical Management at the Management Development Institute, New Delhi. More recently he has joined the Asian Productivity Organization, Tokyo.

Introduction

Biogas technology[1] represents one of a number of village-scale technologies that are currently enjoying a certain vogue among governments and aid agencies and that offer the technical possibility of more decentralized approaches to development. However, the technical and economic evaluation of these technologies has often been rudimentary. Therefore, there is a real danger that attempts are being made at wide-scale introduction of these techniques in the rural areas of the Third World before it is known whether they are in any sense appropriate to the problems of rural peoples.

In response to the interest in biogas and other rural energy systems shown by a number of Asian researchers, the International Development Research Centre commissioned this state-of-the-art review so that it might form a basis of further discussions concerning the direction of future biogas research.

This book, which is divided into three chapters, represents a multidisciplinary approach to the problem and attempts to review existing work rather than to champion particular solutions. The first chapter establishes in broad terms the energy options facing rural communities in the Third World and considers in detail just what is known about the technical aspects of biogas production. This is done by assembling details of known small-scale digester designs and by reviewing the literature to establish the technical parameters determining digester performance.

[1]Biogas technology is based on the phenomenon that when organic matter containing cellulose is fermented in the absence of air (anaerobically) a combustible gas (methane) is formed.

A microapproach to the social and economic appraisal of rural technologies, which stresses both the need to examine technologies in their social context and the need to compare biogas investments with alternative uses of the resources available in specific rural locations, is presented in the second chapter. This approach is discussed in relation to a number of the better attempts that have been made to evaluate biogas.

The third chapter complements the other two by presenting practical field experience. It is based on an extensive survey of a large number of biogas plants and their supporting infrastructure in India, Thailand, Indonesia, the Philippines, South Korea, and Japan.

The technology associated with the production of methane at the village level is in a much greater state of change than is popularly assumed. However, much of the data on both the new systems and the more traditional ones are generally unreliable and inconsistent. It would therefore appear advisable, before any major commitments are made to this form of energy and fertilizer production, that a much more systematic approach be made to research, development, and evaluation.

Many of the technical and economic evaluations that have been carried out so far have been applied to only a limited set of the known techniques, and comparisons have been made between biogas and other systems at the 'high' end of the technology spectrum. Given the current bias in the distribution of the world's research and development effort it is hardly surprising that in these comparisons the underdeveloped small-scale techniques sometimes appear to be inferior. With the fluid state of biogas technology and the unusual interest currently being shown in it, it would seem to be relatively easy to design and build biogas plants that could be operated in rural situations to meet certain social objectives and yet still compete with 'higher' technologies even in conventional terms of profit and capital required per unit of output.

The viability of a particular biogas plant design depends on the particular environment in which it operates. Therefore, the research problem becomes one of providing a structure in which technologists, economists, and users of the technology can combine to produce both the appropriate hardware for various situations and the infrastructure that is necessary to ensure that the hardware is widely used.

Our objective, then, is to stress the need to examine a wider range of technical and economic alternatives for meeting the energy and fertilizer needs of rural peoples. It is our hope that this survey contributes to this process by showing what has already been done, by pointing out pitfalls, and by indicating the major gaps that still remain.

Anaerobic Digestion: The Technical Options

Leo Pyle

Whatever the context of the possible application of biogas technology, choices will always have to be made between alternative courses of action. The choice may be to install a methane generator or to do nothing at all, or there may well be complicated issues of choice between a number of alternative developments. This preliminary section sketches some of the technical alternatives that could be involved and suggests sources for further study.

An effort is made both to put in perspective the state-of-the-art of biogas technology and also to help define the range of system objectives and boundaries relevant to this discussion. Existing digester designs ('core' technology) are therefore described and their features compared and contrasted; this description being later extended to include the 'peripheral' technologies involved in different biogas systems.

The more important aspects and features of the science and technology of biogas — both core and peripheral — are then discussed in order to delineate the more significant ways in which digester behavior can be affected, changed, and measured, and to outline the areas where there are clear gaps in knowledge or where research and development work is progressing, or might be pursued. The problems of evaluating biogas performance, and of deciding where research and development might be justified are also discussed.

It must be emphasized that this discussion is not meant to be prescriptive. An evaluation of the merits of biogas technology, or of the value of investment in research and development, can only be made within a clearly defined set of objectives and within the context of a given technical and social system. This section attempts to be no more than one input to such an evaluation.

The major considerations that are likely to weigh in considering the contribution of biogas to a rural or village environment include its contribution to supplies of fuel, energy, and fertilizer, and to waste treatment, public health control, and sanitation, as well as its use of local resources (material and human).

There will also be a number of wider considerations such as the technology's contribution to indigenous technologies and social and economic development.

Thus we can consider various alternatives, for example: alternative methods of supplying local energy/fuel needs (including methods of production, distribution, and use); alternative methods of using local materials (e.g. cowdung, crop residues, and wastes); alternative methods of supplying fertilizer needs; alternatives in public health control; alternative systems with anaerobic fermentation as the "core," including different end uses; and finally, alternative arrangements/designs of the digester for biogas production. Although there will be some overlap among these different categories, this listing affords a convenient basis for discussion.

Alternative Energy and Fuel Sources

The main sources of energy that could be provided to rural areas are listed in Table 1. This gives a qualitative picture of the substitutability of the different energy sources for household, agricultural, or (small)

Table 1. Main energy sources that could possibly be provided to rural areas.

Energy source	Household			Agriculture/Industry		
	Cooking	Lighting	Heating	Power[1]	Transport	Heat energy[2]
Electricity	(X)	X		X		
Coke, coal	X					X
Kerosene	X	X	X	X		X
Diesel			X	X	X	X
Gas	X	X	X	X	X	X
Wood	X		X			X
Straw, vegetable wastes, crop residues	X		X			X
Dung	X		X			
Solar energy	(X)			(X)		X
Hydro				X		
Wind				X		
Alcohol				X	X	

[1]Includes, for example, pump sets.

[2]Includes steam.

NOTE: (X) represents methods that are likely to be very expensive, be of limited application, or need further development.

industrial use, but gives no indication of the costs or likely appropriateness of the energy sources.

It is also relevant to consider the major existing or potential production methods for the fuels listed, because this consideration can weigh very heavily when considering the choice of technology, and because of wider implications, such as ecological suitability. The primary energy sources can be classified as broadly nonrenewable or renewable: nonrenewable sources include coal, coke, oil, natural gas, and nuclear; renewable include solar, wood, dung, vegetable matter, water, and wind (see Table 2).

Although some of the fuels listed are classified as renewable, this may only be true under controlled conditions. For example, the use of wood as a major source of energy in some developing countries has led to deforestation on an extremely serious scale.

No attempt will be made to describe current patterns of energy use in developing countries: it is sufficient to note the extremely high dependence in rural areas on noncommercial fuels. Many of the methods of use are carried out with very low efficiency or with serious health consequences (e.g. burning of dung indoors creates a source of eye complaints). The questions of energy utilization in the rural areas of poor countries are described and analyzed by Makhijani and Poole (1975) and Makhijani (1976).

The appropriateness of a particular energy source for a given situation depends upon a number of factors, such as the availability of primary resources (e.g. coal, water, etc.) and the economies of scale in production. For example, a possible benefit of using crop residues or solar energy is that these might be feasible on quite small scales and thus be appropriate to rural areas at an early stage of mechanization and development. The practicality of processes centred on biogas production will be the subject of later discussion. First, however, it is instructive to consider alternative methods of gas production.

Alternative Methods of Gas Production

The combustible fraction of biogas is methane; however, there are other possible

Table 2. Some renewable and nonrenewable sources of common energy and fuel types.

End source of energy/fuel	Nonrenewable	Renewable
Electricity	Coal, oil, gas-fired power station, nuclear	Hydroelectricity (solar)
Kerosene, diesel	Oil	
Gas	Natural gas, oil, coal	Celluloses, vegetable wastes, etc. (biomass)
Alcohol	Oil, gas	Celluloses, starches, sugars, etc.

combustible gases such as (mixtures of) hydrogen, carbon monoxide, and higher *n*-paraffins (such as butane and propane).

In industrialized countries methane is much used as a fuel or chemical feedstock. It is a major constituent of natural gas, and can also be manufactured by gasification or reforming from fossil and nonfossil fuels to yield substitute natural gas (SNG). These processes are highly sophisticated technically and are carried out under elevated temperatures and pressures. It seems highly unlikely that they will be serious contenders as technologies appropriate to rural application either in terms of levels of technology employed, capital intensity, or scale of operation. However, for large-scale operations gasification must be considered seriously. Ifeadi and Brown (1975) consider the process promising above 100 tonnes/day (equivalent to the manure from 17000 dairy cows!)

An alternative route, which may in the future be more appropriate, is pyrolysis. In this process the biomass (wood, coal, vegetable, etc.) is heated in an air-lean environment to 400-1000 °C. The products depend on the feed material and the operating conditions, but usually comprise three phases: a solid char, which can be subsequently used as fuel; an oily fraction; and a combustible gas (which may have considerable CO and H_2 content) that can be further processed to produce a gas with a high methane content.

The promise of the process, for relatively small-scale applications, is reviewed by Pyle (1977). For example, Tatom et al. (1975) report a small, transportable pyrolysis unit able to treat wood, peanut shells, sugar waste, trash, etc. The fuel products from the unit are char (with a heating value (H.V.) of 30000 kJ/kg), oil, and gas. Preliminary experience shows that the process is promising for low moisture content wastes where the energy needed to evaporate the moisture is small. However, a good deal of work remains to be done to develop appropriate designs. Makhijani and Poole (1975, p. 100) have commented favourably on the Chinese experience with pyrolysis.

The same range of feed materials can also be fermented anaerobically (at 30-60 °C) to produce biogas. The three methods sketched in Fig. 1 can be seen as potential alternatives within the framework of Fig. 2. And, in fact, whether the wastes are animal or vegetable products, they can be considered to have their basic energy supply in the capture (by photosynthesis) of solar energy (see Table 3).

Figure 3 illustrates this point and shows the potential for gas cleaning and CO_2, water, and nutrient recycling.

Materials Suitability

McCann and Saddler (1976) considered the economics of pyrolysing wheat straw using the Garrett process and found that it yielded between 0.32 and 0.40 tonnes of oil (at a lower calorific value than fuel oil) per tonne raw material. All the gas produced and most of the solid char was recycled to provide heat for the process. Despite this, the preliminary economic analysis of the process (under Australian conditions) looks extremely promising. However, if a wet manure is used to feed the pyrolysis unit, it is necessary to evaporate about 80-85% of the weight before treatment. On the basis of 1 tonne wet matter, containing 200 kg dry matter (calorific value 14500 kJ/kg), one would need to evaporate up to 800 kg of water (i.e. 4 kg water/kg dry matter) at an

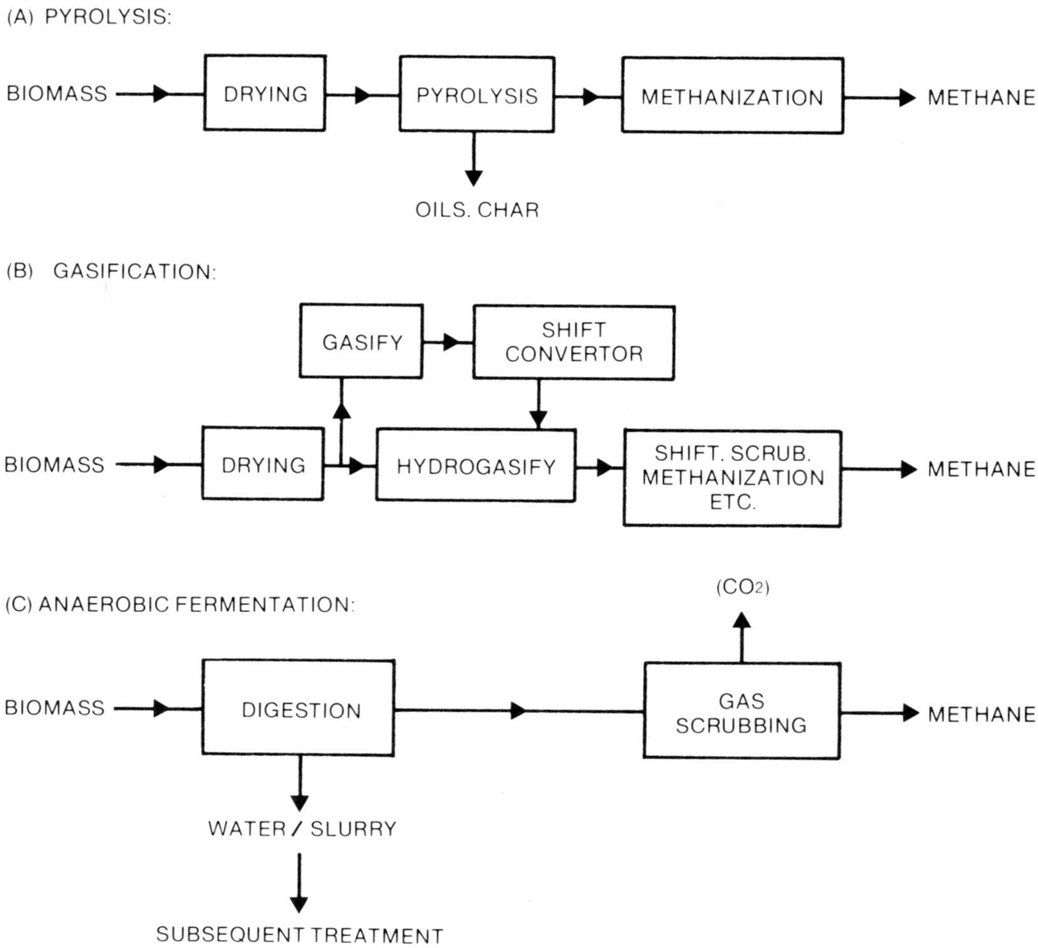

Fig. 1. Three principal methods for anaerobic conversion of feed materials to biogas.

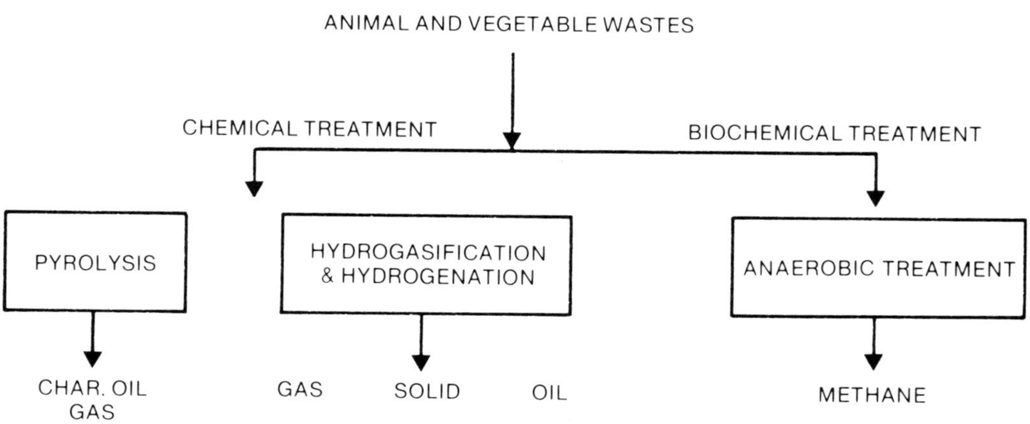

Fig. 2. The three methods of conversion shown in Fig. 1 can be seen as potential alternatives within a framework of energy supply.

Table 3. Examples of fuel-gas production methods from biogas (from Klass 1976).

Biomass	Reaction	Conditions	Products[1]
Pine bark	Pyrolysis	900 °C ambient pressure	low value gas, char, oil
Rice straw	Pyrolysis	200-700 °C ambient pressure	low value gas, char, oil
Cellulosic refuse	Hydrogasification	540 °C, 70 bar	high value gas, char
Wood, paper	Digestion	30 °C, 30 days	intermediate value gas
Grass	Digestion	48 °C, 10-28 days	intermediate value gas
Water hyacinth[2]	Hydrolysis + digestion	48 °C, 28 days	intermediate value gas
Seaweed[3]	Hydrolysis + digestion	33-48 °C, 20-50 days	intermediate value gas
Unicellular algae[4]	Hydrolysis + digestion	35-55 °C, 30 days	intermediate value gas

[1]Low value = 3700-16 500 kJ/m³; intermediate value = 16 500-30 000 kJ/m³; high value > 30 000 kJ/m³.
[2]*Eichhornia crassipes.*
[3]*Macrocystis pyrifera* (giant kelp).
[4]*Scendismus* spp., *Chlorella* spp.

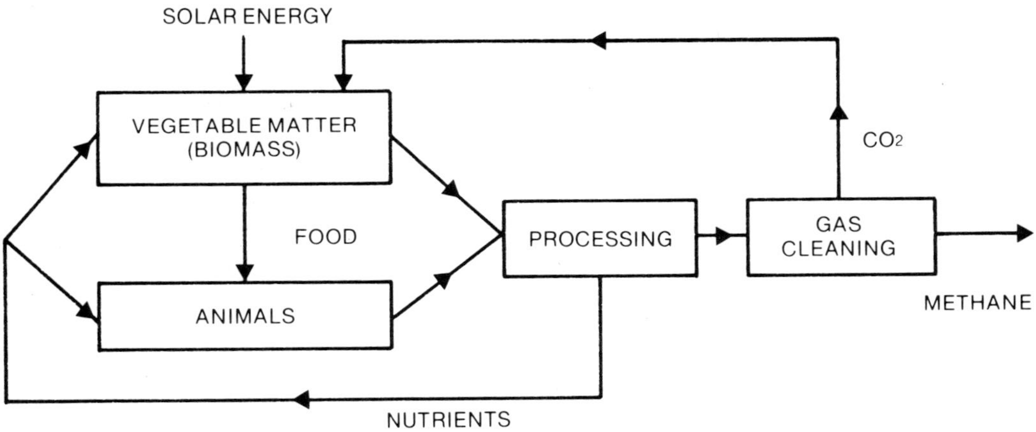

Fig. 3. Wastes, whether animal or vegetable products, have their basic energy supply in the capture (by photosynthesis) of solar energy.

energy cost of approximately 2500 kJ/kg of water or 10 000 kJ/kg dry matter. The feedstock energy (%) used for evaporation is shown, as a function of moisture content, in Fig. 4. Thus for materials of high moisture content there is a strong argument in favour of anaerobic digestion because no evaporation is necessary.

Alternative Methods of Waste and Biomass Utilization

The preceding discussion has touched on production of fuel gas from animal and vegetable wastes, and conceivably, from crop residues. It is, however, necessary to view the question within a wider context in order to establish an optimum strategy for local communities. For example, one has to consider current patterns of usage to ensure that people are not worse off — in material or resource terms — as well as establish alternative methods of use. For example, sugar cane may be used as a fuel, or alternatively, the cane can be used in paper manufacture, which may be more 'profitable.' However, the implications of a switch in end use raise very complex issues. Our purpose is not to oversimplify the issues involved but rather to point out some of the

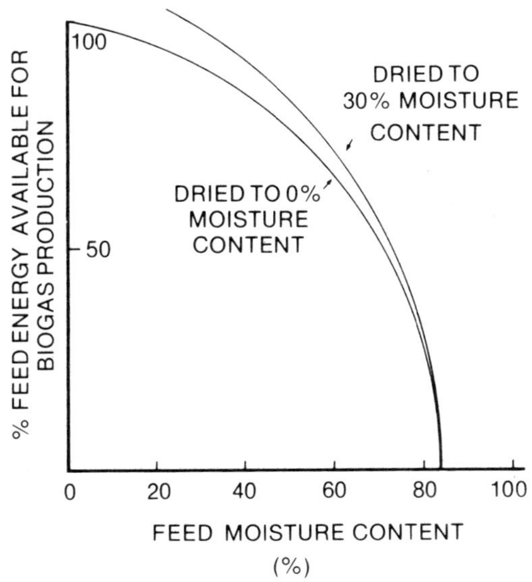

Fig. 4. Effect of feed moisture content on amount of energy available for biogas production.

more obvious choices, because the question of feasibility can only be raised within particular situations and sets of objectives.

Alternatives Based on Animal and Crop Wastes

Biogas production is often suggested in situations where animal wastes are used as a major source of household energy. The potential advantages include: (1) the replacement of an inefficient (but traditional) fuel with a more efficient and flexible one; (2) the recoupment of the fertilizer value of the waste, which is lost if the dung is burned; and (3) the benefits to public health (especially in reducing eye diseases) if the cleaner, less smokey, gas is used. However, the question remains: Does this represent the best use of the waste?

Some of the alternatives for biochemical and chemical processing of raw materials containing cellulose are outlined in Fig. 5.

Composting

The need for organic matter in agriculture is well recognized. In the context of fertilizer shortages and increasing prices, the need to use organic wastes has taken on a new dimension, and their contribution certainly cannot be underrated. For example, it has

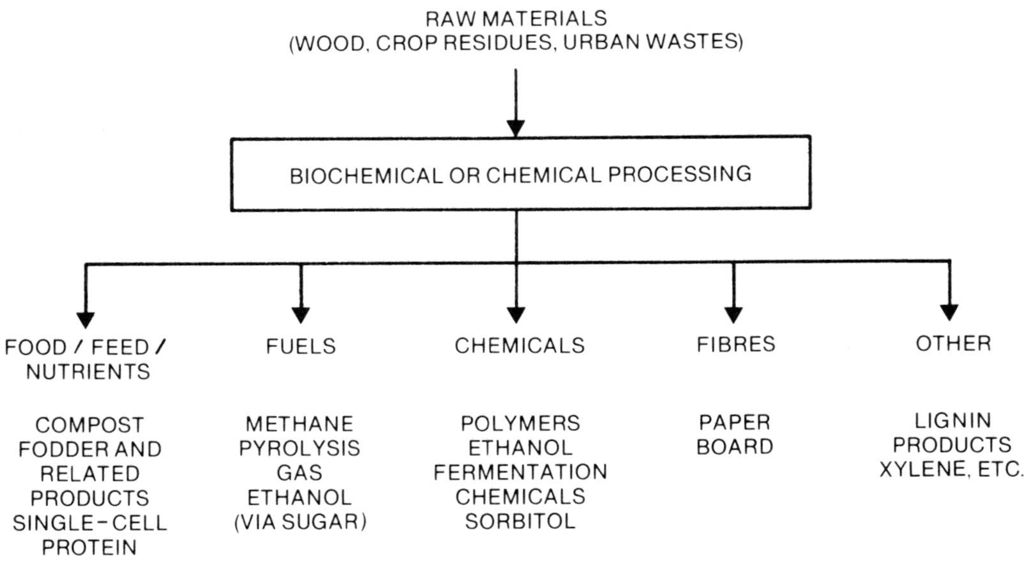

Fig. 5. Simplified product chart based on cellulose.

been reported that 85% of China's cultivated land is treated with organic manures (night soil, compost, green manures, etc.). One major advantage of this type of manuring is its contribution to recycling and conservation of plant nutrients (see Gotaas 1956). There is strong evidence of the need for naturally derived nutrients and humus in situations where heavy reliance is placed on 'chemical' fertilizers (see Dhua 1975).

When dung is used not as manure but as a fuel, a link in the nutrient cycle is broken. This could be avoided if the manure was used as an input for anaerobic digesters. It can be argued that this would be preferable to composting (where the nutrient qualities are conserved, but the fuel value of the wastes is lost). However, one must also consider the zero capital-cost of unmechanized composting, and the more complex issues related to public health (e.g. pathogen destruction).

Although one would like to calculate the 'trade-off' between the net benefits of anaerobic digestion and the fertilizer value of composted materials, it is not possible at this stage to obtain quantitative measures of the fertilizer/nutrient value of the slurry effluent from biogas plants, nor to compare the value with composted materials.

It is also noteworthy that composting, being an aerobic process, is exothermic. Under normal conditions this heat is wasted (it is very low grade heat) except insofar as it is responsible for the destruction of pathogens. However, it may be worthwhile in some circumstances to use the heat from composting material as an energy input to a biogas plant. This can be done by surrounding the biogas plant with compost.

From the point of view of efficiency of converting feed (grass, grain, etc.) into human food, animals leave a good deal to be desired: the major proportion of the feed is converted not into meat protein but into protein in the manure. Reuse of the digestible portion of the feces could lead to significant reductions in the overall cost (and therefore increased efficiency) of raising cattle (Pimental 1975), swine, and poultry. There are a number of ways in which this can be done (Perrigo and Demmitt 1975).

Use of Untreated Material

There have been a number of studies of the use of untreated or slightly treated manures as a ration supplement, and Perrigo and Demmitt report generally favourable results when the use of bovine manures was limited to about 10% of the cow's diet. The practice is currently forbidden in the USA and parts of Western Europe: it is, however, practiced in the U.K., where dried poultry manure is currently being used successfully. Because the costs involved are small, careful attention should be paid to the possibilities of this method.

Manuric Modified Silage

Anthony has successfully developed a process where manure can be incorporated into silage. The manure is mixed with Bermuda grass (in ratios up to 1.3:1) and left to ferment. The 'wastelage' contains 10% crude protein and 60% digestible nutrients (dry basis), and has produced positive results in feeding trials.

Fermentation-Based Processes

Two broad types of processes using microorganisms to increase the nutritive content of manures are discussed by Perrigo and Demmitt (1975). The first involves the use of bacterial cultures to ferment the manure to silage. The second involves culturing the microorganisms on the manure substrate, harvesting them, and finally processing them to increase digestibility. For example, Ward and Seckler (1975) discuss a process where a high-protein fraction from cattle manure is fed directly to poultry. The proposed method involves anaerobic fermentation of the waste followed by fractionation to three products A, B, and C. These three fractions are characterized in Table 4.

Ward and Seckler claim that the high-protein fraction from one dairy cow can support 30 hens. In fact, because poultry manure is high in uric acid, when it is dried it

Table 4. Characteristics of the three fractions derived from fermentation-based processes.

Fraction	Characteristics	Potential use
A	High fibre (equiv. to corn silage)	Feed for cattle, sheep
B	High protein (20-30%)	Feed for ruminants or nonruminants
C	High ash	Soil amendment

is itself a possible feed supplement. Thus, this method offers a potential reduction in the acreage needed for producing animal feed. Although presently unproven, this method should be considered as one alternative among many. With the recent interest shown in processes of this type, considerable improvement may be expected in the near future.

An alternative method of converting wastes into protein is their use as a substrate in algal culture. In principle various wastes can be fed to ponds in which unicellular or filamentous green or blue-green algae are cultured. Yields of approximately 80 000 kg/ha/yr (dry matter) have been achieved under laboratory conditions (Boersma et al. 1975, Pantastico 1976). The efficiency cf these waste ponds depends on numerous factors: the physical conditions (incident light (most algae cannot grow oxidatively on sewage in the dark), pH, temperature, mixing); the quantity and biochemical oxygen demand (BOD) loading of the feed; the pond dimensions; nutritional conditions such as the presence of micro- and macronutrients; and the species present in the population, their frequency of harvesting, etc. (Shaw 1973).

Physicochemical Processes

One of the major limitations on the ease of utilization of cellulose-based materials is their rate of hydrolysis to produce sugars, which can be more directly utilized. A number of physicochemical pretreatments have been proposed to prepare manure and agricultural wastes for refeeding, recycling, or processing. For example, heating under pressure and/or the use of acids or bases (e.g. caustic soda) can be used to improve the digestibility of cattle manures (Stidham et al. 1973; Klopfenstein and Koer 1973). Robb and Evans (1976) report the use of sodium hydroxide in the recovery of nutritive materials from cereal straw. Hydrolysis has also been used to replace the slow systematic breakdown of ligno-cellulosic plant tissues to sugars, for subsequent fermentation to single-cell protein (Worgan 1973).

Curiously, there are few reports of such methods as part of biogas production, where the rate of hydrolysis of the feed is largely responsible for the slowness of the process. This point is considered in more detail later.

The methods discussed above for waste recycling have as a major objective the improvement of the efficiency (viewed in input: output terms) of animals as food producers. The same arguments hold if cattle, for example, are used for work. In other words, the object is to increase the ratio of human food:animal food or work:food, and the feasibility has to be determined by weighing the cost of treatment and recycling against the improvement in efficiency.

As has been seen, it may be advantageous to use the animal waste (whether treated or not) as a food additive for another population.

There are clearly a number of implications of such schemes, and these are discussed below in the context of integrated food, energy, and waste-treatment cycles. Before discussing integrated systems, however, we will consider other possible uses of wastes or biomass.

Other Fermentation Processes

Considerably wider possibilities than those outlined above exist. Three such processes are considered here, and more information is presented in the recent symposium by UNITAR (1976).

Protein production from carbohydrate wastes

Carbohydrates are the largest renewable source of carbon compounds available for conversion into protein for human or animal food. A variety of materials can be used as substrates for the production of edible protein in the form of yeasts (e.g. *C. utilis)*, fungi (e.g. *F. semitectum*), and bacteria (e.g. *Escherichia coli*). However, there is still work to be done on processing methods and on the palatability and acceptability of such products before these processes are likely to be used on a wide scale for bulk food preparation.

Worgan (1973) gives a good review of some of the methods of producing edible protein. Their relative biological efficiency may be judged from Table 5, which is based on using both sucrose and oat husk as carbohydrate sources.

Table 5. Relative biological efficiency of some methods of producing edible protein based on using either sucrose or oat husks as the carbohydrate source (from Tables 7 and 8, Worgan 1973).

	Carbohydrate (g) to yield 100 g protein		Protein doubling time (h)	
	Sucrose	Oat husk	Sucrose	Oat husk
C. utilis	400	666	5	5
F. semitectum	400	910	5	5
E. coli	286	—	5	—
Beef cattle	1900	9224	2800	2800

Relative biological efficiency is also reflected in land productivity calculations: for sugar beet the yield of protein can be 2800 kg/ha/yr; the yield of beef protein is, by contrast, about 42 kg/ha/yr. Of course, cattle are more than just beef producers: their wastes can be utilized and they are essential sources of power for much of the world's population.

Production of glucose, alcohols, etc.

Starches and celluloses can be used to produce sugars. Starch, for example, can be hydrolyzed by acids or enzymes and then fermented to ethyl alcohol. Celluloses too can be broken down in a similar way. Processes for the enzymatic hydrolysis of cellulose to glucose have recently been developed, but the estimated cost is, at present, high (McCann and Saddler (1976) quote a figure of $US 0.22-0.55/kg glucose) and ethyl alcohol made by the route is probably uneconomical (Fig. 5 illustrates the process).

The hydrolysis of starch is much easier than the hydrolysis of cellulose, and McCann and Saddler (1976) claim that alcohol production from starchy substrates (e.g. cassava) is cheaper than from cellulose. The costs of production of various fuels estimated by McCann and Saddler are given in the Table 6. The potential of cassava for fuel, food, and industrial chemical (via alcohol) production is at present a virtually unexplored region.

An excellent discussion of ethanol production by fermentation is given in the review paper by Trevelyan (1975), where both the process and end-use alternatives are discussed in detail.

The different end uses of cellulose can be assessed by comparing the value (market price) of the various products. Table 7 (Dunlap 1975) gives such estimates, but does not take into account the difficulties or costs associated with processing nor the value of by-products (for example, in biogas production the fertilizer value of the slurry is extremely significant).

Alternative Sources of Biomass

Biogas production is usually considered as a method of treating animal and vegetable wastes. However, it is advisable to consider wider possibilities; for example, the feasibility of growing renewable crops for energy production.

Potential raw material can be divided into two classes: (1) land grown; and (2) water grown (fresh or sea). The choice of the most appropriate crop will depend on many technical and social factors. Included in the technical factors are: attainable growth rates under the climatic conditions in question; nutrient and water demands; ease of

Table 6. Costs (calculated for Australian conditions) of some photobiological fuels (from McCann and Saddler).

Fuel	Raw material	Process	Saleable by-products		Comparative cost[1]	Primary energy input	Secondary[2] energy input	Efficiency %	
			As fuel	Other	$/$10^9$J	(MJ/kg product)	(MJ/kg product)	NUEP	Overall
Alcohol	Cassava tops & tubers	Enzyme hydrolysis/ Batch fermentation	—	Fibre (animal feed), fusel oils	8.4	75.1	17.3	17	32
Alcohol	Eucalyptus	Acid hydrolysis/ Batch fermentation	—	—	13.4	42	105	-180	20
Alcohol	Eucalyptus	Enzyme hydrolysis/ Batch fermentation	—	—	20.1	—	—	< 0	—
Methane	Cereal straw	Bacterial fermentation	—	Biomass slurry	4.2	105.9	20.0	34	44
Methane	Eucalyptus	Bacterial fermentation	—	Biomass slurry	5.5	105.9	20.0	34	44
Pyrolytic oil	Cereal straw	Flash pyrolysis (Garrett process)	Char	—	3.3	50.6	4.8	52	58
Pyrolytic oil	Eucalyptus	Flash pyrolysis (Garrett process)	Char	—	4.3	50.6	4.8	52	58

[1]Comparative costs of other energy sources at time of study ($/$10^9$J): Kuwait crude oil (US$10 per bbl) 1.25; syncrude from coal 1.2-1.9; gasoline (70¢ per gallon, taxed) 4.45; diesel fuel (35¢ per gallon, untaxed) 2.0; No. 6 fuel oil ($75 per tonne) 1.7; and natural gas (Cooper Basin) 1.15.

[2]Includes energy cost of harvesting, transport, process fuels, electricity, and ingredients.

Table 7. Average to good annual yields of dry matter (from U.K./ISES 1976).

	t/ha/year	g/m²/day	Photosynthetic efficiency (% of total radiation)
Tropical			
Napier grass	88	24	1.6
Sugar cane	66	18	1.2
Reedswamp	59	16	1.1
Annual crops	30	—	—
Perennial crops	75-80	—	—
Rain forest	35-50	—	—
Temperate (Europe)			
Perennial crops	29	8	1.0
Annual crops	22	6	0.8
Grassland	22	6	0.8
Evergreen forest	22	6	0.8
Deciduous forest	15	4	0.6
Savanna	11	3	—
Desert	1	0.3	0.02

harvesting; water content of harvested crop; amount of harvested feed; and reactivity or biodegradability.

A summary of known average or achievable yields, with some approximate cost estimations (under U.S. conditions), is given in Tables 7-11, and a good discussion of the suitability of the highest yielding of these potential sources is given in Alich and Inman (1975).

Alternative Sources of Plant Nutrients

Nutrients vital for plant growth and development are based on carbon, oxygen, hydrogen, macronutrients (nitrogen, phosphorus, and potassium), and, finally, secondary and micronutrients. In the case of micronutrients, toxic effects from oversupply can cause serious problems.

The first four elements occur in air and water; phosphorus, potassium, and nitrogen 'fixed' by organisms are found in the soil and are subject to exhaustion as they are removed by plants. Their availability depends on environmental factors such as temperature, moisture, and acidity. Soils may be replenished by applying nutrients in the form of organic or 'natural' fertilizers such as: manure, plant residues, composts, animal by-products (blood, bone meal, etc.), (treated) human wastes, and slurry from biogas plants, combined with chemical fertilizers or by chemical fertilizers alone (urea, ammonium bicarbonate, ammonium phosphates, potassium sulfates).

There is a worldwide trend toward the use of high analysis and complex (chemical) fertilizers. Since the early seventies,

Table 8. Photosynthetic yields (mg CO_2 fixed/cm²/h) at normal and enhanced CO_2 gas levels (from U.K./ISES 1976).

	Normal	Enhanced
Maize, sorghum, sugar cane[1]	60-75	100
Rice	40-75	135
Sunflower	50-65	130
Cotton	40-50	100
Soybean, sugar beet	30-40	56
Oats, wheat, barley	30-35	66
Tobacco	20-25	67
Tomato, cucumber, lettuce	20-25	50
Tree species, grapes, ornamentals, citrus	10-20	40

[1]These are the only C_4-species in the table.

Table 9. Some high short-term dry weight yields of crops and their short-term photosynthetic efficiencies (from U.K./ISES 1976).

Crop	Country	g/m²/day	Photosynthetic efficiency
Subtropical			
Alfalfa	US, California	23	1.4
Potato	US, California	37	2.3
Pine	Australia	41	2.7
Cotton	US, Georgia	27	2.1
Rice	S. Australia	23	1.4
Sugar cane	US, Texas	31	2.8
Sudan grass	US, California	51	3.0
Maize	US, California	52	2.9
Algae	US, California	24	1.5
Tropical			
Cassava	Malaysia	18	2.0
Rice	Tanzania	17	1.7
Rice	Philippines	27	2.9
Palm oil	Malaysia (whole year)	11	1.4
Napier grass	El Salvador	39	4.2
Bullrush millett	Australia, NT	54	4.3
Sugar cane	Hawaii	37	3.8
Maize	Thailand	31	2.7

NOTE: Yields in $g/m^2/day$ can be converted to t/ha/year by multiplying by 3.65.

Table 10. Productivity and energy conversion in agricultural crops on an annual basis.

Crop	Country	t/ha/year	Photosynthetic efficiency
Temperate			
Rye grass	UK	23	1.3
Kale	UK	21	1.1
Sorghum	US, Illinois	16	0.6
Maize	UK	17	0.9
	UK	5 (grain)	0.2
	Canada, Ottawa	19	0.7
	Japan	26	1.1
	US, Kentucky	22	0.8
Potato	UK	11	0.5
	Netherlands	22	1.0
Sugar beet	UK	23	1.1
Wheat (spring)	UK	5 (grain)	0.2
	US, Washington	12 (grain)	0.4
	US, Washington	30 (total)	1.1
Barley	UK	7 (grain)	0.3
Rice	Japan	7 (grain)	0.3
Subtropical			
Alfalfa	US, California	33	1.0
Sorghum	US, California	47	1.2
Bermuda grass	US, Georgia	27	0.8
Sugar beet	US, California	42	1.2

(continued)

Table 10. Productivity and energy conversion in agricultural crops on an annual basis *(concluded)*.

Crop	Country	t/ha/year	Photosynthetic efficiency
Potato	US, California	22	0.6
Wheat	Mexico	18	0.5
	US, California	7 (grain)	0.2
Rice	Australia, NSW	14 (grain)	0.4
	US, California	22	0.6
Maize	Egypt	29	0.6
	US, California	26	0.8
Tropical			
Napier grass	El Salvador	85	2.4
	Puerto Rico	85	2.2
Sugar cane	Hawaii	64	1.8
Oil palm	Malaysia	40	1.4
Sugar beet	Hawaii (2 crops)	31	0.9
Cassava	Tanzania	31	0.8
	Malaysia	38	1.1
Sorghum	Philippines	7 (grain)	0.2
Maize	Thailand	16	0.5
	Peru	26	0.8
Rice	Australia, NT	11 (grain)	0.2
	Peru	22	0.7
Rice + Sorghum (multiple cropping)	Philippines	23 (grain)	0.7

Table 11. Biomass yields and estimated costs (from Saddler et al. 1976; Klass 1975).

Material	Yield (t/ha/year)	Estimated cost
		($/tonne)
Eucalyptus	16	1.70
Cassava: tops	12	1.70
tubers	17.5	2.10
Kenaf	30	2.60
Elephant grass	68	1.10
Sugar cane	44	1.1-1.4
		($/10 BTU)
Corn	2.27	9.7
Corn silage	15.9	1.31
	6.5	1.88
Conifer	—	1.25-1.75
Poplar	10	0.9-1.0
Sugar cane	25	0.63
Kenaf	20	0.61
Kenaf	6	1.4
Land/water based	20-50	0.4-1.5

however, escalating prices and supply shortages due to the increasing costs of raw material and energy have affected both production and freight costs. The international price of ammonia, for example, increased nine-fold between 1972 and the end of 1974, and led to declining demand, especially in developing countries. At the same time, experience has shown that fertilizer supply is perhaps the single most important technical factor in agricultural growth. The need for increases in productivity per hectare is increasingly important and calls for an acceleration in the supply of fertilizers. The extent to which chemical fertilizers will supply this need will depend on factors that vary from country to country: distribution and credit facilities; availability of appropriate supplies; supply of complementary inputs; know-how and skills to run production facilities near capacity; and technical backup to the farmers.

Organic Fertilizers

Recent practice has tended toward monocultural production based on chemical fertilizers. There is plenty of evidence, however, that biologically supplied or recycled organic nutrients (composting, biological nitrogen fixation, nutrient recycling) represent an alternative to reliance on chemical inputs.

It is impossible to generalize on the potential contribution from wastes because their composition varies enormously. Taiganides and Hazen (1966) provide typical NPK contents, and Ames (1976) gives some average figures for a range of organic manures (Table 12). An approximate idea of the potential contribution of animal manures can be seen from Tables 13 and 14.

On average 75-85% of the major nutrients and 40-50% of the organic matter in the feed are present in the manure. Urine contains 40-70% of the fertilizer value of the manure. Taiganides and Hazen calculated the potential annual value of the manure per 1000 lb (450 kg) live weight to be $71 for poultry, $42 for hog, and $26 for cow manure (in 1966 US dollars), valued at the cost of equal amounts of commercial NPK.

Pantastico (1976) outlines the elements that are relevant to an assessment of the role of organic nutrients, and gives data on the composition of manures and composts of different types (compost, stable manure, night soil, raw straw, plant ash, and green manure). It has been estimated that wastes from animals, plants, and humans could supply developing countries with six to eight times more nutrients than they derive from chemical fertilizers. These figures can be compared with estimates of the NPK concentrations of fertilizer and manure in the

Table 12. Average chemical composition (%) of some organic manures (from Ames 1976).

	N	P	K
Bulky organic manures			
Farmyard manure	0.5-1.5	0.4-0.8	0.5-1.9
Compost (urban)	1.2-2.0	1.0	1.5
Compost (rural)	0.4-0.8	0.3-0.6	0.7-1.0
Green manures (various averages)	0.5-0.7	0.1-0.2	0.8-1.6
Edible oil cakes			
Coconut	3.0-3.2	1.8-1.9	1.7-1.8
Cotton seed (decorticated)	6.4-6.5	2.8-2.9	2.1-2.2
Cotton seed (undecorticated)	3.9-4.0	1.8-1.9	1.6-1.7
Groundnut	7.0-7.2	1.5-1.6	1.3-1.4
Manure of animal origin			
Dried blood	1.0-1.2	1.0-1.5	0.6-0.8
Fish manure	0.4-1.0	3-9	0.3-1.5
Bird guano	7-8	20-25	2-3
Bone meal (raw)	3-4	20-25	-
Bone meal (steamed)	1.0-2.0	25-30	-
Activated sludge (dry)	5-6	3-3.5	0.5-0.7
Settled sludge (dry)	2-2.5	1-1.2	0.4-0.5
Night soil	1.2-1.3	0.8-1.0	0.4-0.5
Human urine	1.0-1.2	0.1-0.2	0.2-0.3
Cattle dung and urine mixed	0.60	0.15	0.45
Horse dung and urine mixed	0.70	0.25	0.55
Sheep dung and urine mixed	0.95	0.35	1.00

Table 13. Average daily manure production and composition of hens, swine, and cattle (from Tables 2 and 3, Taiganides and Hazen 1966).

	Hens (1.8-2.3 kg)	Swine (45 kg)	Cattle (450 kg)
Wet manure (kg/day)	0.1	3.2	29.0
Total solids (% wet basis)	29.0	16.0	16.0
Volatile solids (% dry basis)	76.0	85.0	80.0
Nitrogen (% dry basis)	5.6	4.5	33.7
P_2O_5 (% dry basis)	4.3	2.7	1.1
K_2O (% dry basis)	2.0	4.3	3.0

that about 16% of the N present in the digested sludge is present as dissolved ammonia, which evaporates on standing (see also Idnani and Varadarajan 1974). The proportion of nitrogen as ammonia varies with the feed: for rice straw the loss is only 8-10% (Acharya 1958).

Where the fertilizer (i.e. nitrogen) content of the slurry is important, it is essential to minimize volatilization losses by using proper storage and application methods. Tanks or lagoons are perhaps the most satisfactory storage method; whereas, loss during application can be minimized if the sludge is injected below, rather than spread on the soil surface.

Table 14. Major fertilizing elements per 450 kg live animal weight (from Tables 2 and 3, Taiganides and Hazen 1966).

	Hens		Swine		Cattle	
	kg/day	kg/yr	kg/day	kg/yr	kg/day	kg/yr
Wet manure	25.4	14600	31.7	10160	29.0	9340
Total mineral matter	1.77	635	0.82	270	0.91	360
Organic matter	5.53	2000	4.26	1540	3.72	1360
N	0.42	150	0.23	84	0.17	63
P_2O_5	0.31	115	0.12	50	0.05	19
K_2O	0.15	54	0.22	78	0.14	51

USA (Table 15), which show that manures generally have a relatively low nutritive value (but of course the organic matter itself serves a vital function).

Other possible sources of fertilizer (Briones and Briones 1976) are industrial wastes (e.g. mud press from sugarcane mills), bean meals, and garbage.

Biogas Plants as a Source of Fertilizer

Several authors have commented on the fertilizer quality of digester slurry (for example, Acharya 1958). Unfortunately data on the fertilizer value of the slurry from biogas plants are inadequate. Theoretically, little of the NPK fed to the digester should be lost during the process because the only loss of N is as gaseous ammonia and this is small during digestion. Acharya points out

Table 15. Estimates of the quantity and value of nitrogen fixed by some of the principal legumes.

	Quantity of N fixed in a growing season (kg per hectare)[1]	Value as fertilizer of N fixed ($ per hectare)[2]
Lucerne (alfalfa)	50-460	15-138
Clovers	50-670	15-200
Other temperate pasture legumes	20-200	6-60
Tropical pasture legumes	20-400	6-120
Peas	30-140	9-42
Chickpeas	100	30
Soybeans	40-200	12-60
Peanuts	70-240	21-72

[1]Ranges of values in recent findings — due to the complexities of the factors involved these ranges are wide, with extreme values in both directions relatively unusual.

[2]The nitrogen fixed was valued conservatively at a 1975 'farm gate' cost of $0.30/kgN.

Other plant nutrients may be conserved and made more available if plant residues are recycled through a digester. The result of using slurries from anaerobic digesters on the land is much the same as using any other kind of compost: the humus material plays a vital role in improving soil properties and texture.

Generally, 50-70% of the degradable organics fed to the digester are decomposed, and only a small proportion (10-20%) of the carbon is converted to cellular matter. Thus, problems arising from using the sludge on land are much smaller than those from using aerobically treated wastes because the smaller quantity of bacterial matter minimizes both smells and insect development.

Biological Nitrogen Fixation

Despite the ever-increasing use of chemically fixed nitrogenous fertilizers it has been estimated that biological fixation contributes about four times as much nitrogen to the soil. One example is the formation of nodules on the roots of various legumes (pulses, beans, nuts, peas, some clovers) by bacteria. Bacteria associated with the roots of nonlegumes (rice, sugar cane, etc.) are also important, especially in the tropics. In addition, other bacteria and blue-green algae (Pantastico 1976) can contribute to the process. The possible contribution of some legumes to the supply of N fertilizer is given in Table 16.

Not all the nitrogen fixed finds its way into the soil (which points to the urgent need for research work), but rotation of legumes and cereals gives a substantial increase in cereal yields. On the other hand, some legumes make very high phosphorus demands and economic responses are achieved only by fertilizing with superphosphate.

The enormous range of options in the provision of chemical fertilizers is too wide for discussion here, but two further comments are perhaps in order. First, the technologies for nitrogenous fertilizer production are marked by considerable apparent economics of scale, as a consequence of their high capital intensity. A methodology for comparing such widely differing alternatives as large-scale capital-intensive ammonia production and smaller, more labour-intensive, technologies (e.g. biogas) is discussed in the next chapter. It should be kept in mind that problems have also been encountered due to either high transport costs or to operation at much less than full capacity. Second, the existence of an 'anti-chemical' lobby must be acknowledged. It is, however, not our intention to take one side or another in this debate: the relative lack of attention to chemical fertilizers must not be taken as implying any prejudice on the writer's part.

Public Health, Waste Treatment, and Pollution Control

A major source of concern throughout the world is the safe treatment and disposal of wastes. Here we are primarily concerned with the disposal of organic wastes, many of which are likely to be degradable. These

Table 16. Average oxygen demand data for farm-animal wastes (from Taiganides and Hazen 1966).

Animal size (kg)	Vol.solid (kg/day)	BOD (kg/day)	BOD (kg/kg VS)	COD (kg/day)	COD (kg/kg VS)	BOD/COD (%)
Hens (1.8-2.3)	0.025	0.008	0.320	0.026	1.04	30.8
Swine (45)	0.43	0.15	0.349	0.57	1.32	26.3
Cattle (450)	3.7	0.58	0.156	4.76	1.29	12.2

wastes are derived from a wide range of sources: animal manures and wastes; farm wastes including residues; night soil; contaminated waste waters; industrial wastes, including wastes from agroindustry; and general solid wastes, garbage. Wastes range in nature from solids to liquids, but our main interest is centred on paste, semiliquid, or slurry-type wastes.

Among the public health considerations that dictate the need for waste disposal facilities are: the transmission of disease vectors, pathogens, etc. (and especially fecal-borne diseases), and associated contamination problems; the problems caused by disposing untreated wastes on land or water (odours, insect colonies etc.); the potential of using valuable materials, if they can be recovered or recycled; and regulations (if any) covering the quality of discharges.

The range of choices available will broadly be: to do nothing; to treat at the source (e.g. composting latrine); or to treat centrally (e.g. community waste treatment facility).

Industrial waste disposal and control cannot be dealt with here although many of the treatment policies will be applicable. Useful background to the use of anaerobic fermentation in controlling industrial effluents can be found in Fair et al. (1966) and Mosey (1974).

Pollutant Strengths

There are a number of indexes of the pollutant strength of an organic waste. One is the biochemical oxygen demand (BOD), which measures the oxygen demand that would be exerted on a water body if the waste were discharged into an (aerobic) water course. It is usually measured at 20 °C over 5 days (hence BOD_5). The chemical oxygen demand (COD) can also be used as an index, but the test does not differentiate between biologically degradable and inert matter. Under some conditions, BOD and COD of a waste will correlate; more often, the correlation is not good. If the tests correlate, the COD measure is superior because it is both faster and more accurate. In assessing animal and human wastes these measures are best expressed in terms of BOD (or COD) (as kilograms oxygen demand) per kilogram of volatile solid (VS) in the waste. Tables 17 and 18 give average BOD data for farm animal wastes and mean figures for BOD and COD for animals and humans. The COD/kg live weight is about the same for hens, pigs, and cattle; the BOD of cattle waste is considerably lower, probably because of the larger proportion of cellulose (which is attacked slowly). The population equivalent (PE) is the number of humans to produce the same daily quantity of waste measured as BOD. In calculating the load on a treatment system, the liquid quantity discharged is also very important. The wide variability (reflecting feeding patterns, health) in the data cannot be overemphasized: the figures quoted are only a guideline.

Treatment Methods

The earliest and simplest form of handling wastes, the cesspool, developed into the septic tank, in which the detectable oxygen level is zero and conditions are anaerobic.

Table 17. Mean BOD and COD of farm-animal wastes (Taiganides and Hazen 1966).

	Live weight (kg)	BOD (kg/day)	COD (kg/day)	PE (BOD basis)
Man (excrement)	68	0.054	—	0.61
Man (total)	68	0.091	—	1.0
Hens	45	0.154	0.526	1.7
Swine	45	0.154	0.567	1.7
Cattle	45	0.059	0.476	0.7

Table 18. Pond characteristics (from Dugan and Oswald 1968).

Pond type	Maximum loading rate (kg BOD_5/ ha-day)[1]	Detention time (days)	BOD removal (%)
Aerobic	22.4-224	30	80
Facultative	22.4-168	90	85+
Anaerobic	112-1120	10	70

[1]i.e. per hectare pond surface.

Perhaps the cheapest and simplest type of device available is a pond (Fig. 6). The zones that are illustrated can coexist in a single pond or exist as separate pond types. The organic matter is converted into: a gaseous product; algae, which can be harvested, discharged, or allowed to settle; and organic volatile acids, methane, carbon dioxide, and stabilized sludge. The reactions in the anaerobic part are thus essentially the same as those in an anaerobic digester.

The optimum growth conditions for the phases in a facultative pond differ. The minimum temperature for effective operation is about 15 °C, and in most climates about 4 m will be the maximum depth for an unheated pond. Odours are caused if the volatile acid concentration becomes too high.

One potential advantage of anaerobic over aerobic treatment systems is that they can be loaded more heavily. The figures in Table 19 allow a rough comparison of ponds loaded with waste water and suspended solids.

Ponds must be carefully designed; for example, in shallow ponds algae can form a blanket giving excessive fatty acid generation even when the (obligate) bacteria needed for methane generation cannot function because of the oxygen present. Aerobic ponds should be less than 0.3 m deep; facultative or anaerobic ponds need to be at least 2 m deep. The range of possibilities for relatively simple treatment systems is shown in Table 20.

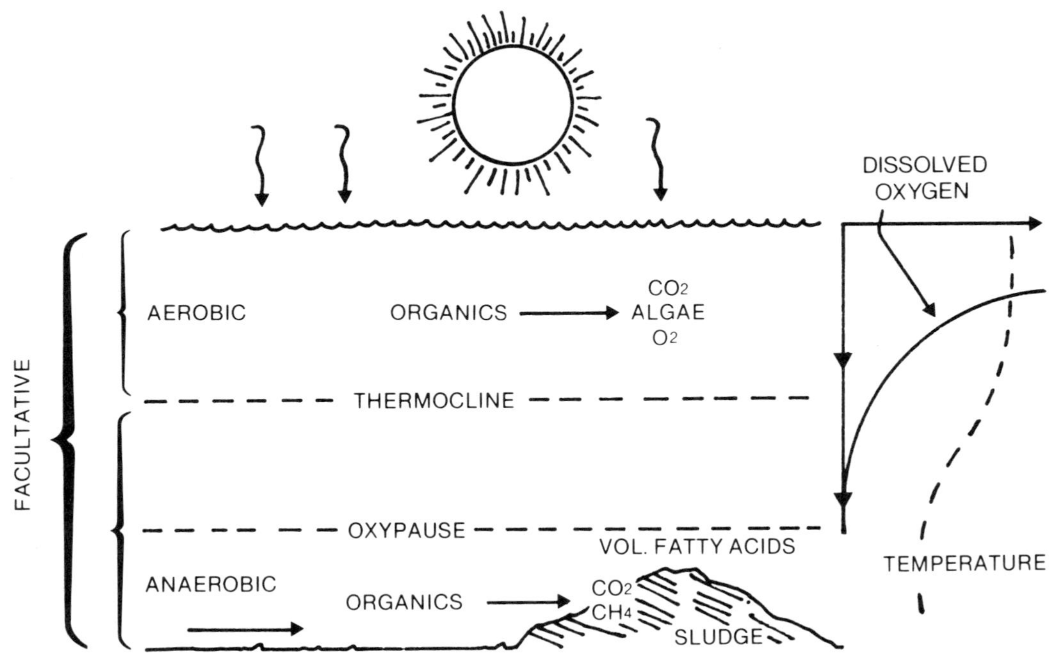

Fig. 6. A pond is perhaps the cheapest and simplest type of treatment device. This profile indicates the zones that may exist within a single pond.

Table 19. Enclosed animal-waste treatment systems (from Loehr 1971).

(1)	+ Water → Holding tank → Land disposal	Cheap, simple
(2)	+ Water → Pond → Land disposal (oxidation, aerated, ditch)	Needs space; better odour control than (1)
(3)	+ Water → Anaerobic unit (uncontrolled, high load) → Aerobic unit → Land disposal	Can handle highly concentrated wastes; secondary (aerobic) treatment necessary
(4)	In-house oxidation or Holding unit → Land disposal	Built into animal house; low handling problems; no need for excess water; semi-solid wastes to dispose of
(5)	Separation at source → Solids to land / Liquids to treatment	Allows treatment adapted to waste
(6)	Drying, incineration	Expensive if mechanized; drying gives solid fuel or fertilizer; nutrient losses
(7)	Composting	Possible handling problems

Pescod (1971) notes that there are relatively few waste-water treatment systems; waste solids (night soil) are often collected, but the operation of septic tanks, cesspools, etc. leaves much to be desired, and the resulting sludges must be treated carefully. Pescod discusses the anaerobic digestion of partially stabilized wastes from septic tanks. In the tropics this is generally not necessary, but his results indicate the technical feasibility of anaerobic digesters fed with night soil or sludge. Treatment is possible at loading rates of up to 4.5 g VS/day/litre digester volume — some three times the normal loading rate for primary sludge treatment. Pescod advocates the use of lagoons for drying stabilized sludge cake, but also cautions of the dangers of open lagoons in the tropics citing the growth of mosquito larvae (*Culex pipiens*) on the surface of the supernatant liquor. It is impossible to disagree with his conclusion that there is a dearth of published information relating to developing countries and that a good deal of research and development is necessary to develop rational design criteria. Jewell (1975) presents a wide-ranging discussion of methods and possibilities in animal-waste treatment with a primary emphasis on intensive agriculture.

The use of manure and composted farm and animal wastes is an important alternative technology that has the advantage of requiring little capital investment. Properly handled, manure has great value both as a source of nutrients and organic matter (Gotaas 1956, the journal *Compost Science*, Klausner et al. 1971, Ames 1976).

One of the major problems in both urban and rural communities is the safe disposal of human wastes. Integrated sewerage schemes are costly. Septic tanks have certain advantages, but they require quite large quantities of water and are unsuitable for areas with high groundwater levels, poor percolation properties, etc. The 'pit' toilet is a sanitary device that is effective if properly located (away from houses, rural areas), constructed, and maintained: it can, however, be a source of odours and groundwater contamination. Recently, there has been a good deal of interest in simple 'composting' toilets that do not need any water and from which the waste products can safely be used as fertilizer. Typical examples of such latrines are the Farallones

Table 20. Summary of operating conditions of a range of different digester designs (1 ft³

Name of plant	Rated capacity	No. of stages	Digester size: Vol. I	Vol. II	Total Vol	Temp (°C)	Cows No. (Qty)	Pigs No. (Qty.)	Chickens No. (Qty.)	Other No. (Qty.)	Vegetable Qty.
CHAN TYPE		1	54 ft³	—	54 ft³	36 no control		1000 lb/d			
CHAN TYPE		1 plus lagoon	3000 l + 3000 l sed.	— —	3000 l	35 no control		230 kg/d			
D.C.S. NEPAL	100 ft³ (winter)	1	8.5m³	—	8.5m³	25	5 (45 kg/d)	—	—		—
D.C.S. NEPAL	180 ft³ max	1	8.5m³	—	8.5m³	about 30	7/8 (60 kg/d)	—	—	—	—
FISHER	100 ft³	1	100 ft³			30	4				
FRY	7000 ft³					30 + control	130 cows or 800 pigs				
KHADI	4m³	1	7m³	—	7m³	30-35 no control	8 (80 kg/d)	—	—	5 people	—
KOREA	?	1	5.5m³	—	5.5m³	24	6 approx (60 kg/d)	—	—	—	Straw
LAPP	?	2	185m³	?	185m³	no data	—	1000 (3000 kg/d)			
LAPP	?	2	5.6m³	?	5-6m³	35	—	no data			
MANN	?	1	465 ft³	—	465 ft³	no data	←	mixed feed		→	
MAYA FARM	—	1 32 X batch	600m³ total		600m³	?	—	5000 (6-8 ton/d)			
NEERI	—	1	230 ft³ (6.5m³)	—	6.5m³	19.7, 14.1, 16.4, 16.1 } water temp.				night-soil digester	
NEERI		1	1m³		1m³	?	no data				
PATEL	100 ft³	1	110 ft³ (approx 3m³)			mesophilic	4-5				
PATEL	300 ft³	1	approx 9m³			mesophilic	12-15				
SCHMIDT-EGGERS-GLÜSS	—	2			18000 ft³ 30000 ft³	30	280 280	←	(or equivalent)	→	no data (Straw)
SCHMIDT-EGGERS-GLÜSS	—	2			15000 ft³	35-40	130				
SINGH	various	←	see various papers by Singh			→					
TAIWAN	2.5m³	1	5.4m³	—	5.4m³	30(?)	—	10-15 hogs plus water (17 kg/d)			
TAIWAN	2.5m³	1	5.4m³	—	5.4m³	30(?)	—	10-15 hogs plus water (17 kg/d)			
TAIWAN	2.5m³	1	5.4m³	—	5.4m³	30(?)	—	10-15 hogs plus water (17 kg/day)			

= 0.028m³, to convert lb/ft³/day to kg/m³/day multiply by 16.02).

Total qty.	Dry solids	Vol. matter	Water	% solids (dry)	Loading rate	Gas product (daily)	BOD In	BOD Out	VS In	VS Out	Residence time	Comments	Reference
	65 lb/d	52.5 lb/d		6.5%	0.97 lb VS/ft³/d	130 ft³			25% destruction		3-4 days	Calculated	—
230 kg/d			10:1		7 kg/m³/d	300 l/kg = 15000 l/day		no data			1 day	Design data	Richard
45 kg/d	18%	6.75 kg/d	1:1	7-9%	0.8 kg/m³/d	100 ft³ (3m³)		no data			approx 60 days	Design data	Finlay (1976)
60 kg/d	18%	10 kg/d	1:1	7-9%	1.2 kg/m³/d	180 ft³ (5m³)		no data			approx 30 days	Design data	Finlay (1976)
40 kg/d	18%	5 kg/d	1:1 (?)	7-9%	0.6 kg/m³/d	100 ft³ (claimed)		no data			no data, approx 30 days +	Design claim	Fisher (1972)
1300-2600 kg/d	18% max	200-400 kg/d	?	8%	6 kg/m³/d	7000 ft³		no data			30 days +	Design claim	Wright Rain (1963)
85 kg/d	18-20%			9%		4m³/d		no data			42 days		Florida (1974)
100 kg/d (fresh)	20% +		no data		3 kg/m³/d	2m³/d (3.3m³/d at 35°C)		no data			20 days	Tentative figures	
3000 kg/d	no data	14%	no data		2.4 kg/m³/d	200m³/d					30 days		Lapp (1974)
—	—	14%	no data		1.6 kg/m³/d	2m³/d?					20 days		Lapp (1974)
300 ft³	no data	no data	no data	no data	no data	1/3 digester volume/day					50-70 days	Batch	Mann (1962)
no data						Approx 300m³/d (1/3 vol/day)		no data			Batch cycle		Obias IHT/BG/(2)/7
					0.104 kg/m³/d 0.076 kg/m³/d 0.096 kg/m³/d 0.086 kg/m³/d	1.2 1.6 (volumes 1.3 per day) 1.4					24 22 26 28	Trade-off effect	Sathian-athan
8.0 kg/d 16.0 kg/d 24.0 kg/d 32.0 kg/d					1.17 kg/m³/d 2.4 kg/m³/d 3.8 kg/m³/d 5.3 kg/m³/d	0.22 0.42 (volumes 0.52 per day) 0.47					50 25 17 12	Trade-off effect	Sathian-athan
50 kg/d	18%		1:1		2.5 kg/m³/d	100 ft³ nominal					25-30 days		"Biogas"
120-170 kg/d	18%		1:1		2.5 kg/m³/d	300 ft³ nominal					25-30 days		"Biogas"
30 kg/d Total X 250		840 kg/d	?		0.105 lb/ft³/d 0.0625 lb/ft³/d	10000 ft³ 14000 ft³							Sathian-athan
					0.06 lb/ft³/d	17300 ft³						Strong temp. effect	Sathian-athan
				7-9%	5 lb/ft³/d	Claims 1/2 to 1 volume per day					30-60 days		Singh (1971)
17.4 kg/d	30% T.S.	4.25 kg/d			0.8 kg/m³/d	0.825 m³/day	about 60% reduction				6 days	See Paper	Chung Po (1974)
17.4 kg/d		4.25 kg/d			0.18 kg/m³/d	2.3 m³/day	about 60% reduction				8 days	See Paper	Chung Po (1974)
17.4 kg/d		4.25 kg/d			0.8 kg/m³/d	3.5 m³/day	about 60% reduction				16 days	See Paper	Chung Po (1974)

composting privy (van der Ryan 1976), the Toa-Throne (de Jounge 1976), or the similar Clivus system, which is essentially a container (able to hold some 4-months waste) with provision for controlled airflow through the aerobic composting heap. Care is needed to ensure that the high temperatures for destruction of the harmful microorganisms (i.e. parasite eggs, protozoa, and viruses) can be attained uniformally. In cold climates, this may well be a serious drawback. The latrines cited are relatively expensive (the Toa-Throne costs up to $1000, the Farallones latrine is quoted at around $100), but cheaper versions are available. The returns to such an investment are partly indirect (improved health, etc.) and partly direct (provision of a fertilizer substitute).

Details of conventional waste disposal methods are to be found in the standard texts on sanitary engineering; most are designed to handle relatively dilute wastes aerobically. More substrates can be handled aerobically than anaerobically, but the contact between oxygen and the substrate has so far limited aerobic systems to dilute, largely stabilized, materials with solids present as fine suspended particles. In systems where solid wastes are degraded aerobically, the main problem is to maintain the solids in a relatively dry, loose matrix. The large energy releases associated with aerobic processes enable a high proportion of the substrate carbon and nitrogen to be converted to microbial cells, themselves subject to further microbial attack, so that disposal of the wet sludge can sometimes lead to further pollution.

There are two major potential advantages of anaerobic digestion systems. First, the possibility of stabilizing waste material for subsequent safe disposal, and collecting the gas produced during digestion. Second, it should be possible to treat more concentrated slurries than with aerobic systems.

In practice, as noted above, the process is not foolproof. Even under controlled operation the slurry may contain considerable fatty acids, ammonia, and other nitrogenous compounds. Further treatment may thus be necessary before the liquid reaches acceptable levels for discharge. Anaerobic systems can certainly treat more concentrated wastes. COD levels of 4000 mg/litre or BOD's of approximately 100000 ppm are possible (Hobson et al. 1974); and solids concentrations between 2 and 10% are normal, as compared with $< 1\%$ in the case of aerobic treatment processes. Anaerobic processes are generally considerably slower, which may lead to cost problems. Anaerobic systems can also handle a range of organic solids (Klein 1972).

Pathogen Destruction

The choice of technology also relates to the possibility of disease transmission. Here we are primarily concerned with fecal-borne disease.

The diseases for which causative or vector organisms are associated with fecal wastes fall into four broad groups, according to the vector: viruses, e.g. poliomyelitis, hepatitis, gastroenteritis; protozoa, e.g. amebic dysentery; bacteria, e.g. typhoid, paratyphoid, dysentery, cholera, TB, enteritis, salmonellosis; helminths, e.g. roundworm, pinworm, sheep liver fluke, bilharziasis.

The hazards associated with treatment processes that handle human excreta thus depend on the incidence of the various organisms, their survival rates, and their subsequent viability in secondary treatment, storage, and discharge to the land.

Most organisms are destroyed during aerobic composting (Gotaas 1956) if temperatures exceed 60 °C for longer than 0.5-1 hour. Most resistant are the eggs of *Ascaris lumbricoides* (roundworm). These can survive 14 days at 35 °C. Low temperature processes must be evaluated carefully (there are a number of WHO documents that cover this field).

Anaerobic digestion (above 30-35 °C) is as good as any other practical treatment process for human excreta. However, there is little documentation dealing with outbreaks of disease in relation to the use of night soil and other wastes on the land. Detailed results relating to anaerobic digesters are discussed more fully later.

Alternative Systems Based on Anaerobic Digestion: Digester Designs

Two broad objectives that are often associated with anaerobic digestion are: the treatment of wastes prior to disposal; and the generation of methane. These are by no means the only objectives; nor are they of comparable importance. The design of digesters reflects the need to meet these objectives. In this section, the main design variants are outlined and a preliminary attempt is made to compare the performance of different designs despite the lack of data.

Feed Materials and Measures of Concentration

Laboratory studies have shown that a wide range of organic matter is biodegradable by anaerobic fermentation. However, one single design is unlikely to cope adequately with all possible substrates because of the different physical conditions and properties and different rates of fermentation that are involved. Designs have been reported that deal with: soluble industrial wastes; sewage and human wastes; animal manures (for which many designs are available); and vegetable and general farm wastes.

The more soluble and easily degradable the substrate, the more easy is the design and operation. In particular, pig wastes are considered to be relatively easy to handle (Hobson et al. 1974); whereas, a good deal of trouble has been reported with straw bagasse or other low density vegetable matter. Such materials very easily form an impenetrable scum on the surface of the digester. However, there is considerable advantage in fermenting vegetable matter because the potential methane production rate per unit mass is higher than that from cow manures that have, in effect, already been through an anaerobic digester (i.e. the rumen).

Concentrations may be measured in various ways, more or less appropriate to digestion. In the case of contaminated water the most usual measures are the BOD, COD, VS, or total organic carbon. BOD levels of up to 10^6 ppm may be treated by anaerobic fermentors. In practice, the concern is more likely to be with solid-bearing inputs such as manures. Slurries with solid contents of about 10% are likely to be quite paste-like, and designs to handle concentrated slurries efficiently would minimize water requirements and the required digester volume.

The most frequently used measure of the biodegradable proportion of the feed is the VS content, which is a close approximation to the potential substrate; with a mixed substrate any overall measure is likely to be over-simplified. The conditions in the digester itself will depend on both the concentration and mean retention time of the feed. In the case of a continuous digester of volume V, feed rate Q(m^3/day), and feed concentration C(kgVS/m^3) loading rate can be defined as:

$$LR = \frac{CQ}{V} \text{ kgVS/m}^3 \text{ digester volume/day}$$

$$= \frac{C}{\theta}$$

where θ = hydraulic retention time.

This is often used as a basis for comparing different digesters. If one digester operates at a higher loading rate than another then either it can process a greater quantity of substrate in the same retention time or handle the same quantity in a smaller time or in a smaller volume.

Batch Versus Continuous

A summary of the operating conditions of a range of designs is given in Table 20, which serves as the basic reference for this section.

Digesters can be broadly divided into either batch or continuous flow. In a batch operation, the raw materials (substrate) are charged into the digestion vessel and the fermentation process can be considered in three stages.

In stage I, the bacterial population begins to establish itself, and following a lag (perhaps many days) gas evolution begins. The gas is likely to be unusable (or even dangerous) as a fuel, with a high concentration of hydrogen sulfide. In stage II (2-4 weeks) the gas production rate increases, passes through a maximum and then begins to decrease. In stage III the gas production rate falls off gradually. The total time for virtually complete digestion is about 60-90 days.

In a continuous operation, the substrate is fed to the digester continuously, so that, once the operation is established the rates of gas production, and input and output are steady with time (see Fig. 7).

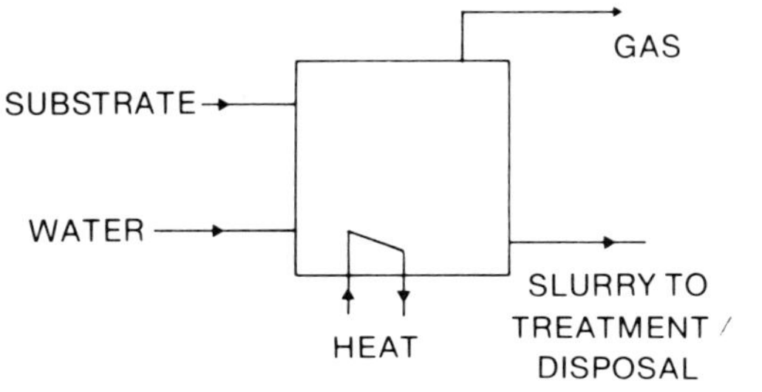

Fig. 7. Schematic process of a continuous operation. Once established, the rates of gas production and input and output are steady with time.

The majority of digester designs are intended for continuous operation. It is often claimed that continuous digestion is more efficient (i.e. has higher gas production rates per unit digester volume) than batch operations. There is, however, little direct evidence of this, and given the relatively high reported failure rate of simple digesters this assertion may be wrong.

In fact, one advantage of a batch operation is that daily attention is not as crucial as with continuous operation where the maintenance of steady operating conditions is vitally important. It is possible to obtain an approximately constant rate of gas production by having a small number of batch digesters connected to one manifold serving a central gas storage facility. One of the more successful installations is at the Maya farm in the Philippines (Obias 1975) where 32 batch digesters, based on swine manures, are in operation. An early European design (Lessage and Abiet 1952) was a batch process, and Mann (1962) has recorded a number of other simple designs.

Mixing

The degree of mixing varies considerably in both batch and continuous processes. The simplest situation is zero mixing. Some simple batch fermentors are of this type, but they are quite inefficient (e.g. the early versions of the 'Ducellier' type were simply loaded with prerotted manure and allowed to digest over a long period). Other batch digesters are mixed by a centrally located stirrer (for relatively dilute wastes), or in the Schmidt and Eggersglüss-type design, by using a pump to circulate the liquid manures.

Mixing reduces stratification and thereby improves contact between the organisms and the substrate; it has also been suggested that mixing increases the rate of decomposition by releasing small trapped gas bubbles from the microbial cell matrix (Finney and Evans 1975), but there is no direct evidence for this. In the absence of mixing the material stratifies as sketched in Fig. 8.

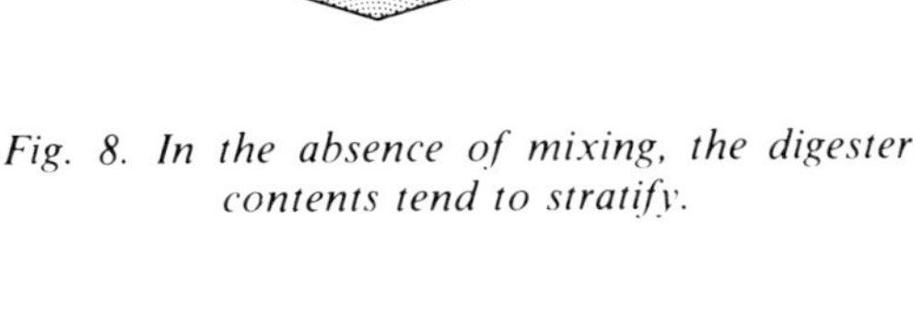

Fig. 8. In the absence of mixing, the digester contents tend to stratify.

Stirring also breaks the scum layer, which if undisturbed can lead to inefficient digestion and can even provide a seal on the digester. If buoyant vegetable matter (e.g.

straw) is to be digested mixing is vital (cf. *Trends in Technology*, Sept. 1974, p. 3).

The simplest types of unstirred continuous devices can be extremely inefficient, especially if no attempts are made to reduce liquid bypassing. With long retention times (>30 days) and low loading rates (<1.6 kg VS/m^3 digester/day) the digester, which is little more than a septic tank, will perform more or less adequately. Many designs, and more recent designs in developing countries, are of this type (see Table 20 for operating data). Operating efficiencies (i.e. rates of gas production/unit volumes of digester) are low ($\ll$1 volume/day).

Despite their simplicity, care must still be taken if the units are to operate successfully. Some typical sources of trouble that are common to many designs (but which can be remedied by careful design and construction) are: blocking of inlet/outlet pipes (remedy: omit all bends, constrictions); leakages to surrounding land (remedy: careful construction, favoured by cylindrical designs); and gasometer toppling/jamming (remedy: design guides correctly) (see e.g. Sathianathan 1976).

If the vessel is stagnant then undigested substrate (especially leaves and large particles) will collect, leading to an interruption of the process. Some designs are particularly prone to this; for instance, rectangular designs and possibly the design proposed by Chan, and Richard (1975). This exemplifies the possible advantage of batch processing at simple levels of technology, for batch digesters can handle a wider range of substrates (e.g. chopped vegetables) than the corresponding simple (unstirred) continuous digester (cf. Mann 1962, p. 240). The gas output from a batch digester can be made approximately constant either by operating single units in rotation or by connecting a series of digesters, each operated out of phase, to a central gasometer.

Rectangular digesters of the 'Chan' type probably have a rather nonideal internal fluid flow pattern. Some designs explicitly aim to eliminate mixing and Fry's (1974) design is perhaps the best known example. Here the attempt is to attain "plug flow" with (inevitably) a degree of stratification.

These designs are fairly simple to engineer and manufacture, and what little data there are suggest that they are not too inefficient. Fry's design is primarily for animal (especially swine) manures but could possibly be adapted for vegetable matter (especially if the digester is at a slight angle to the horizontal). Fry (*New Alchemy Newsletter* no. 3, 1973) discusses how the design copes with scum accumulation by using a small drag device. Some gas storage space is provided above the digesting materials but an additional gasometer is necessary.

Other simple designs with little or no agitation appear to operate continuously without problems from accumulating sludges. The KVIC designs from India (Sathianathan 1976) are examples of successful design. Some modifications of the septic tank, although not efficient as gas producers, allow regular and continuous removal of settled sludge, which has not happened in the septic tank. Figure 9 shows

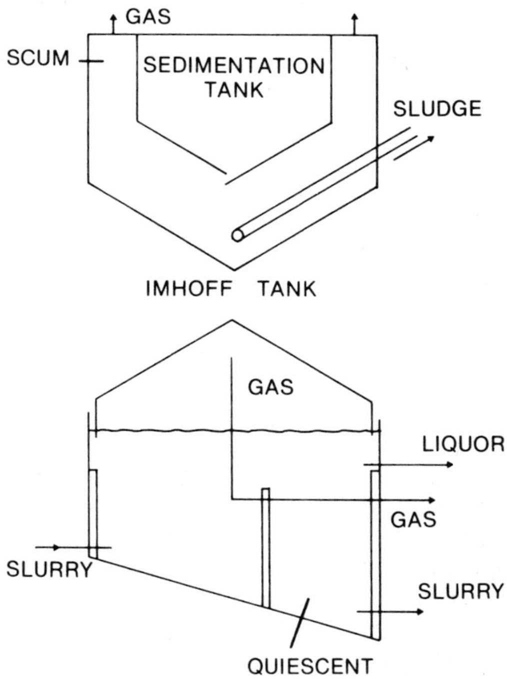

Fig. 9. A modified Imhoff tank and a second type of sludge digester.

a modified Imhoff tank and another type of sludge digester.

The second of these designs incorporates a simple sedimentation/separation volume, (this is a feature of many sewage waste treatment systems and of the two-stage designs proposed by Ram Bux Singh 1971 and Sathianathan 1976).

It is not clear whether the larger KVIC biogas plants — with a dividing wall in the digester (see Fig. 10) — operate as two-stage

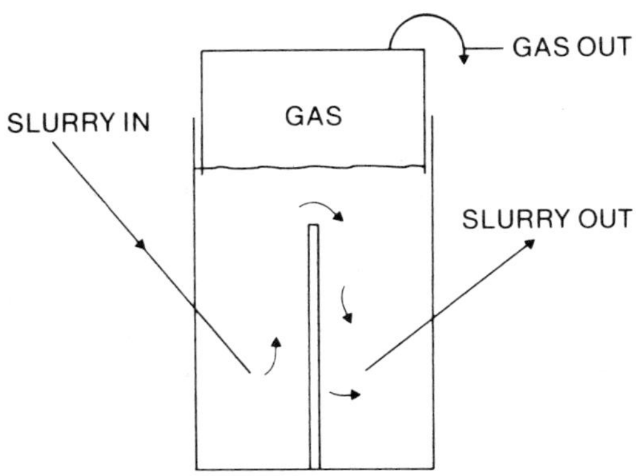

Fig. 10. Simplified diagram of the KVIC digester.

devices; there is some mixing, and it is reported that performance depends on the digester depth or the length/depth (L/D) ratio. Presumably the 'correct' choice of L/D ration ensures reasonable mixing (e.g. by gas bubbles).

That mixing improves the rate of digestion has been shown in full-scale and laboratory-scale studies. Current practice in developed-country applications is toward continuous stirring (as opposed to 20 minutes or so per day in the Ram Bux Singh designs). Mixing may be by stirrers or agitators (hand/cattle powered for intermittent stirring; continuously powered for constant stirring), jet pumps (Ram Bux Singh 1971), or by pumping the digester contents in a recycle. This can be coupled with external heat exchangers (See Meynell 1975).

Such well-stirred digesters are known as 'high-rate' digesters and have been successfully used with urban sewage (Meynell 1976), and various animal wastes (Hobson et al. 1974). Their efficiency is seen in their ability to handle large input flows (loading rates up to 10 times those in conventional digesters can be achieved; typical improvement is two or three times). (Typical loading rates: conventional: 0.6-1.6 kg VS/m^3/day; high rate: 2.4-6.4 kg VS/m^3/day.) A qualitative comparison of the digester types outlined above is given in Table 21 (from Meynell 1976, p. 47).

Quantitative Comparisons of Digester Efficiencies

Unfortunately there is little reliable data for a serious comparison of digester efficiencies. Table 20 represents a modest attempt to collect typical data, but it must be emphasized that much of the data on 'simple' digesters operating in developing countries is speculative and perhaps hopeful rather than realistic. In many cases it is impossible to assess how closely plants approach their design or claimed performance. Some of the data are of extremely doubtful quality (cf. Malynicz' discussion of a paper by Chan, Univ. Papua New Guinea 1973).

The wide range of operating conditions and efficiencies is shown by the data in Table 20. In general, retention times of 30 or more days are typical of 'conventional' poorly mixed digesters; 10-20 day retention times are possible for high rate digesters; and Hobson et al. (1975) have shown that stable digestion can be achieved with continuously loaded digesters (on pig waste) at 2-3 days retention time, although performance begins to fall off below about 10 days. Overall reductions in total solids of approximately 40%, volatile fatty acides and BOD of about 90%, and COD of 40% are typical. These figures can be compared with the data of Chung Po et al. (1975) in an exemplary study of a simple digester (also using pig waste) (Table 22).

Some data are plotted in Fig. 11A. But, there is currently insufficient data to com-

Table 21. Comparison of different types of digester systems.

Digester type	Suitable wastes	Volumes, solids content	Typical retention times (days)	Degree of mixing	Operating temp. (°C)	Gas production	Degree of control required	Comments
Batch	Agricultural, Irregular or seasonal, fibrous or difficult to digest	Low volumes up to 25% solid	60 or more	Little needed	Usually 30-35	Irregular and discontinuous	Little once started	Messy and time consuming to start
Plug-flow Horizontal Vertical	Agricultural, continuous or regular flows, less fibre content.	Larger volumes 5-15% solids	30-60	Occasional	30-35	Continuous	Simple	Loading and scum removal can be messy
Conventional sewage works	Continuous sewage sludge	Less than 5% solids	30-60	Occasional	30-35 or unheated	Continuous	Simple	Not very effective
High rate sewage digestion								
Primary	Sewage sludge	4-10% solids	10-30	Regular	30-35	Continuous	More sophisticated	Automatic
Secondary	From primary digesters	4-10% solids	20-60	None	Unheated	None collected	Simple	
High Rate	Agricultural industrial	4-15% solids	5-20	Continuous	30-35	Continuous	More sophisticated	Can be automated
Anaerobic contact	industrial (agricultural)	Low solids	0.5-5[1]	Continuous	30-35	Continuous	Sophisticated	Automatic
Anaerobic filter	Industrial	Low solids (low organic contact)	0.5-5[1]	None needed	30-35	Continuous	Sophisticated	Automatic

[1]Liquid retention time.

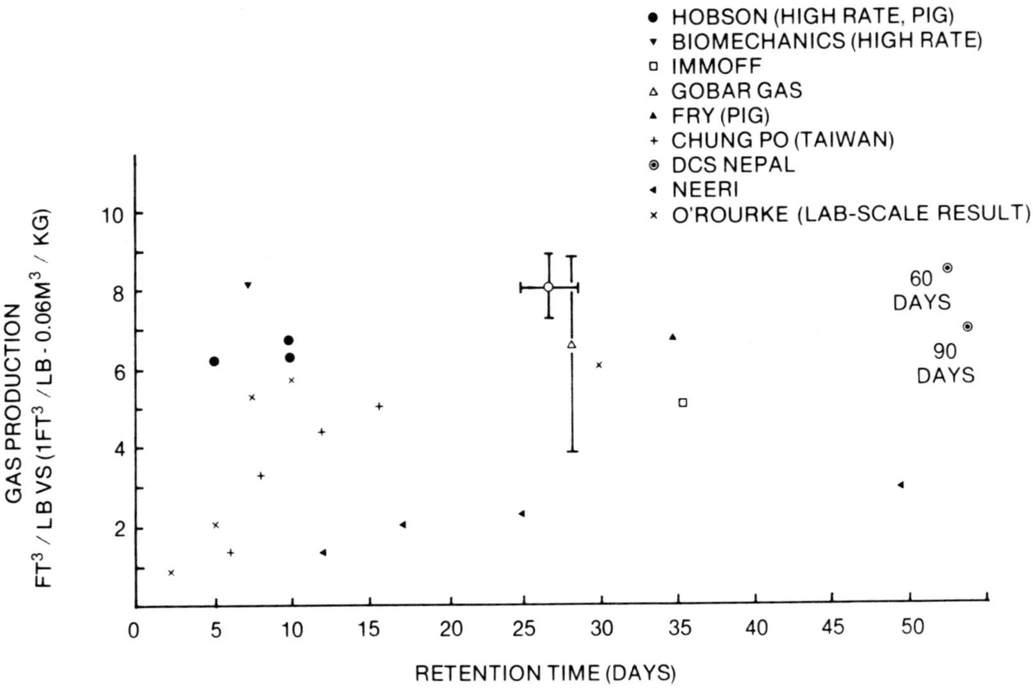

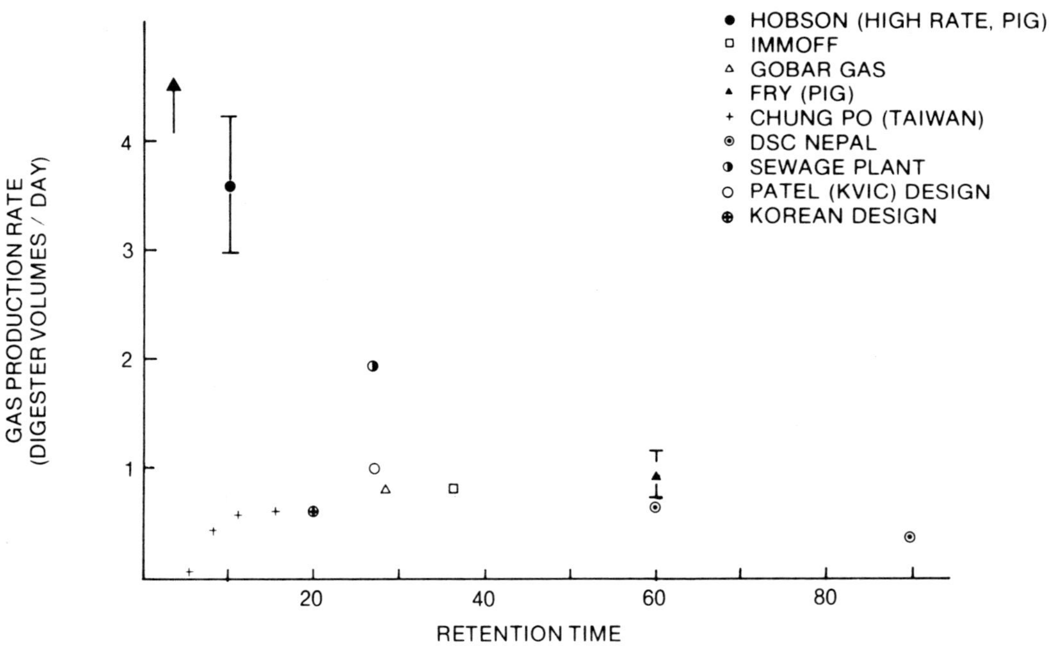

Fig. 11. A very wide range of operating conditions exists. Here, reported gas production rates are plotted against retention time.

pare the performance of different digesters and different levels of sophistication.

The wide range in operating conditions is however shown by Table 20 and Fig. 11B. Retention times range from as low as 2-3 days (with soluble wastes) to 60-70 days in the case of a Fry digester. Loading rates from 0.5 (sewage works) through 2.9 gobar gas) to as high as 11 kg VS/m^3/day are reported. The efficiency of gas production in digester volumes produced per day varies from around 0.50 (Imhoff, Gobar Gas, Fry) to as high as 4 (Rowett Research, Hobson et al.)

Table 22. Results obtained with a simple digester fed[1] with pig waste (Table 3, Chung Po et al. 1974).

Retention time (days)	4	8	12	16
l gas/kg TS destroyed	—	803	1001	993
l gas/kg VS destroyed	—	822	1032	1019
l gas/kg COD destroyed	1201	1318	1501	1455
l gas/kg BOD destroyed	2944	3222	3930	3890
TS reduction (%)	—	56.6	61.1	66.7
VS reduction (%)	—	66.9	71.5	78.5
COD reduction (%)	49.2	57.4	67.7	75.8
BOD reduction (%)	62.6	73.4	80.4	86.7

[1]Loading 0.768g VS/litre/day.

Cell Holdup

The process schemes illustrated in Fig. 7 rely on the establishment in situ of a viable acclimatized microbial population. Bacterial growth rates under anaerobic conditions are slow and thus, in the case of continuous operation, it is possible that the throughput of material could be fast enough to remove bacteria as quickly as they are able to form (i.e. 'wash out'). Even with long retention times the population densities of the bacteria may be so low as to limit the fermentation rate. Various schemes have been devised to attempt to overcome this by using two stages so that the two main bacteriological processes can operate under more favourable conditions (e.g. Ghosh and Pohland 1974). So far these ideas have not been demonstrated with full-scale plants using animal or vegetable waste feeds. Another scheme is to settle out some of the microbial particles (which are attached to the solid substrates) after the digester, and then to recycle either a fraction of the sludge and/or the microorganisms; the net effect is to increase the microorganism concentration in the digester. This type of 'contact' process is illustrated in Fig. 12.

It has been found necessary in some cases (e.g. meat-packing wastes) to use vacuum

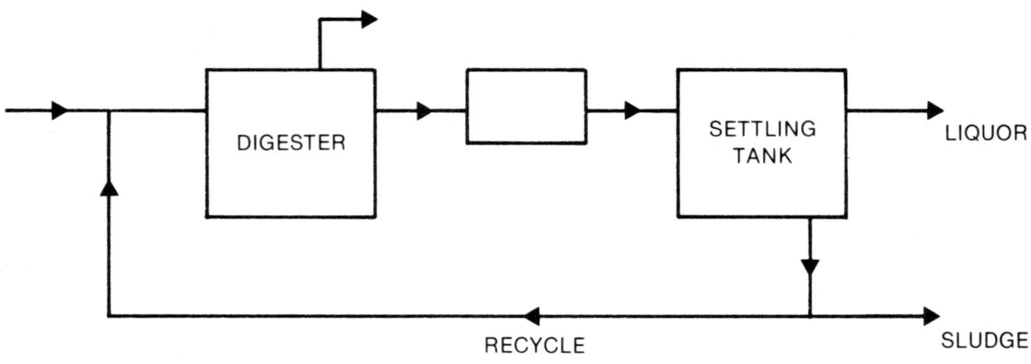

Fig. 12. A 'contact' process, which recycles some of the sludge, has the net effect of increasing the microorganism concentration in the digester.

degassing to remove dissolved and entrapped gases (McCabe and Eckenfelder 1953, vol II (Steffen A.I.)). Torpey and Melbinger (1967) found that the required digester volume was reduced by two-thirds, indicating that with recycling the potential for capital savings is substantial. There is considerable room for study of this type of development in relation to farm wastes. An alternative for increasing the microbial holdup or concentration is the 'anaerobic filter' in which the wastes pass through a bed of concentrated solids, perhaps immobilized on inert particles. This has only been demonstrated on low strength soluble wastes, but extension to higher solids concentrations should be possible.

Temperature and its Control

The digester temperature is a key variable in determining the rate of fermentation. In rough terms, fermentation starts at about 10 °C, and increases rapidly with increasing temperature up to the normally recommended operating temperatures of 30-35 °C. At higher temperatures (50-60 °C) thermophilic bacteria take over and the rates are substantially higher (e.g. McCarty (1964) quotes relative rates of 1.9:1 at 55 and 35 °C).

Anaerobic digesters are extremely sensitive to fluctuations in temperature (Trevelyan 1975). Catastrophic failures can result from temperature changes of only a few centigrade degrees caused by the addition of cold-water feed, a rapid drop in external temperature, etc. Thermal-stability control is therefore very important.

The process only generates a little heat, which is insufficient to keep the temperature at 35 °C if the input and surrounding temperatures are more than 5 °C less. There is thus a need for good insulation and (in colder climates) for some method of heating.

Many simple digesters (Chan, KVIC (gobar), Fisher, etc.) are, like the septic tank, poorly insulated and unheated. In winter or at night, temperatures inevitably fall and so does production. Some more sophisticated designs have heating and insulation (cf. Ram Bux Singh 1971). Although some designs rely on capturing incident solar heat, this is a small input if the heating surface is a (painted) gasometer. The heat demand of the process is shown schematically (but not to scale) in Fig. 13, and possible heat economies, or utilization methods, are shown in Fig. 14.

Heat losses from a below-ground digester are not zero even in hot climates. There are no theoretical problems in design

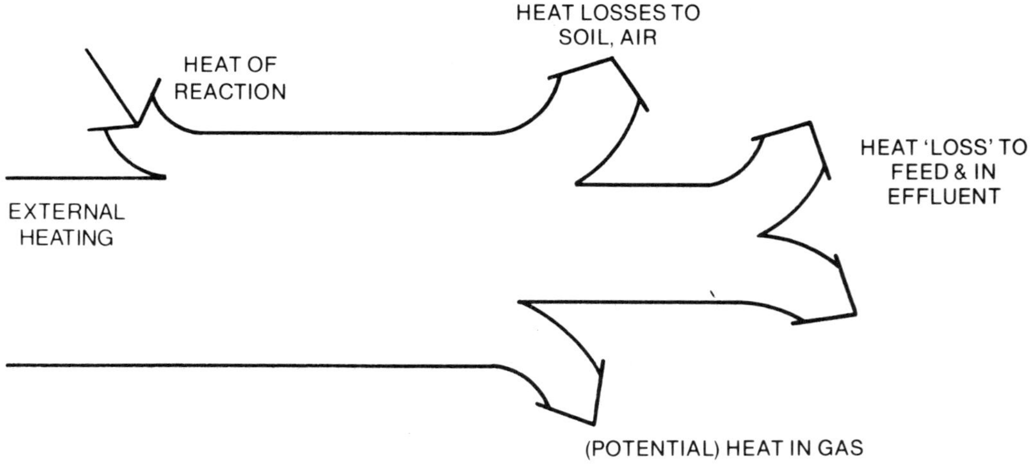

Fig. 13. Heat demand of digestion process (not to scale).

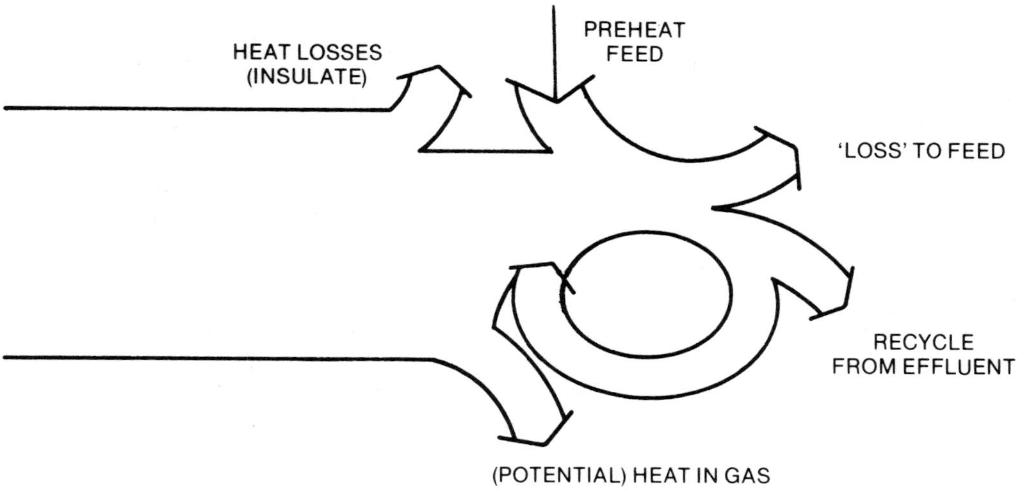

Fig. 14. Possible heat economies, or utilization methods, that may be employed in the digestion process.

calculations: heat transfer coefficients, thermal properties, etc. are all known sufficiently accurately. It may also be possible to save on gas use by using solar heaters either to preheat the feed or to run a simple hot-water circulation loop as a thermosyphon.

Lesage and Abiet (1952) provide a comprehensive discussion of insulation and simple heating systems. One should minimize the digester surface area (cylindrical section being the best basis). Lesage and Abiet also describe an ingenious heating system in which the digester is surrounded by a thick compost layer (the heat liberated during aerobic composting is high and although 'low grade' is well-suited to economizing on thermal energy utilization).

High-rate digesters are almost invariably heated. In practice, internal coils or an external heat exchanger, through which the digester contents are pumped, are used. The latter method is efficient and avoids the problems of scale formation and local overheating near the internal tubes.

It is generally accepted that operation under thermophilic conditions is rarely worthwhile when handling rural waste inputs. But laboratory results (Cooney and Wise 1975) suggest that the question may be open, and Japanese experience using soluble industrial wastes is better under thermophilic rather than mesophilic conditions. Clearly there is a trade-off between capital savings and increased efficiency at higher temperatures, and the costs (in use of gas to heat the digester) of maintaining the temperature. The location of the optimum will depend on local factor prices and should be examined carefully before automatically assuming that operation at 35 °C is 'best.'

Many problems in operating digesters derive from operation with low surrounding temperatures. A simple inventory of the heat loads on the digester (Fig. 13) will give a useful guide to the most likely sources of energy conservation.

Batch reactors are also subject to temperature fluctuations but not to regular disturbance from cold feed. The use of simple but uncontrolled external heating by an aerobic composting pile may be especially beneficial.

Loading Rates

The loading rates of digesters vary widely (Table 20). The advantages of operating at high concentration (loading rates) are to: minimize digester volumes at the same overall residence time; cut down heat load on system; and reduce water requirements and water disposal problems.

A major problem in dealing with very concentrated wastes (greater than 10% solids) is handling the very stiff slurry. It may, however, be possible to operate at higher input concentrations. Some of the early batch systems reviewed by Tietjens (1975) used concentrated loadings and Wong-Chong (1975) discusses operation at 20% dry solids content. This could give a volume reduction of up to 50% over a conventional digester and reduce the problems of disposal/treatment of the supernatant liquor. Problems were encountered due to ammonium buildup (toxicity) with high protein content wastes; clearly a good deal of work still remains to be done.

There is a difference between the simple gobar gas deep-well Indian designs and the rectangular section designs of Chan and Richard. The Indian plants operate at high solids loading and use much less water than the rectangular designs. There is also no separation between the slurry and supernatant liquid. The units seem to be quite well mixed and have operated for many years without any sludge buildup. The rectangular designs (claimed to be designed for detention times of about 1-2 days, which seem impossibly low for proper operation) consume large quantities of water and are very troublesome in that there is a need to remove accumulated sludge at regular intervals.

Unfortunately there are not enough data available for a comparison of the economics of high rate/high loading systems with simple digesters.

Feed Composition

Most continuous digesters are designed to handle animal wastes. Various studies have been made (Acharya, Idnani and coworkers) to investigate the effects of adding small quantities of other organic matter in an attempt to increase the digester efficiency. Sathianathan (1975, p. 38) for example, implies that it is useful to add nitrogen (in leguminous plants) to accelerate the fermentation of cow dung, which may be limited by nitrogen availability. Other authors (e.g. Finlay 1976) mention the possible advantages of adding small quantities of urea; others recommend the use of urine. Careful laboratory studies (c.f. Idnani's work) show that the effects of mixed substrates are not simple.

The ability of batch reactors to handle vegetable wastes has been mentioned. However, there is little guidance on which to base a precise evaluation of the possibilities of handling largely vegetable feeds. Laboratory and pilot-scale studies have shown that grass, (Boshoff 1965; Hadjitofi 1976), coffee bean wastes, etc. may be fermented, yet it is reported (Anon. 1976) that digesters are not able to handle bagasse, coir, and insoluble cellulosic materials. In many situations pretreatment of the digester feed (chopping, soaking, etc.) may be necessary, but there is little data available to enable the degree (or cost) of these operations to be specified. Ram Bux Singh (1971) mentions the problems of scum formation and the flotation of buoyant vegetable matter to the surface (taking with them attached microorganisms).

Obviously, the operating conditions (crudely measured by the carbon/nitrogen ratio) must be maintained in the desired range for fermentation to proceed. Assuming that this is so, Sathianathan quotes production rates of biogas per kilogram dry matter for a range of substrates (and gives data for the digestion of mixtures of night soil/manures, Table 23). These figures show the relative disadvantage of cow dung against other substrates.

Table 23. Production rates of biogas per kilogram of dry matter (from Sathianathan 1976).

	Production rate (m^3/kg DM)
Pig manure	3.6-4.8
Cow manure	0.2-0.3
Chicken manure	0.35-0.8
Sewage	0.35-0.5
Straw, grass	0.35-0.4
Green vegetables/wastes	0.35-0.4

He also quotes figures that show a decrease in total gas production and gas production per kilogram of VS added as the concentration of solids in the feed increases. It is not clear, however, if these data refer to constant detention times. Chung Po et al. (1974) show an increase in efficiency as the retention time increases at constant substrate feed rate (i.e. decreasing inlet concentration): the economic optimum as a function of concentration, loading, residence time, and substrate material is clearly well worth further study (see Appendix 1 for a detailed list of operating data).

Many authors suggest the use of 'starters' or 'seeds' by adding discharged slurry from one digester to promote another (e.g. Sathianathan 1975, p. 41; Alicbusan 1976). This scheme is similar to the cell-recycle schemes discussed earlier. For batch and continuous fermentors improved rates of gas yield have been reported, which is hardly surprising because an acclimatized population and partly digested slurry is being recycled. Care must be taken to avoid the buildup of toxins at high recycle rates.

In view of the interest in using effluents as a source of nutrients for algae, water hyacinths, etc. (Prasad et al. 1974), there are few studies of digester behaviour when the feedstock is comprised of algae or water hyacinth. There is an urgent need for further detailed study of such processes (see for example *Chemical and Engineering News*, 22 March 1976, p. 23; Wolverton and McDonald 1976).

Engineering Design, Construction Materials

The broad design categories have been discussed. One factor that could dramatically change the economics of biogas generation would be a sharp reduction in the capital cost of the digester (see Barrett 1978; ICAR 1975; Prasad et al. 1974; Sathianathan 1975).

There are two possible digester types: those with an integral gasometer and those that feed a separate gasholder. In both cases the gasholder itself is usually an inverted metal cylinder or 'box' that is free to move vertically, so that the pressure inside the gasholder remains constant (the pressure is determined by the weight of the holder and its cross-sectional area). The advantage is that the gas supply pressure remains constant (at a few inches water) and gas supply to the consumer is steady and controllable.

Possible disadvantages are: price (the gasholder is the most expensive part of a gobar gas unit (ICAR 1976); corrosion resulting from the acidic conditions and the H_2S inside the digester and corrosion on the outside lip of the holder (regular painting and maintenance is needed to remedy this); and problems with gas offtake — many systems employ a flexible gas offtake via the top of the gasholder, but the pipes can crack and give serious operational problems. One remedy is to take the gas via a fixed pipe taking care that liquid does not get into the pipe. The corrosion problems of a separate gasholder are less severe than with an integral holder.

Alternatives have been suggested for the metal gasholder. One possibility is to use a wooden/bamboo framework covered with plastic. It may also be possible to use ferrocement. Another possibility is the 'neoprene bag' digester under trial in Taiwan (Chung Po et al. 1974). The digester is made of 0.55-mm hypalon laminated with neoprene and reinforced with nylon sheet. The digester and gasholder can be combined in one bag, so the potential cost of the digester becomes extremely low.

Digesters can be made from relatively cheap local materials (stone, mortar, cement), but must be constructed extremely carefully to avoid leakage. One ingenious possibility is a digester design of Chinese origin presently being promoted in Pakistan (Appropriate Technology Development Division, Govt. of Pakistan 1976). The principle of this unit, which can handle animal manures and some vegetable wastes (especially as 'accelerators'), is shown in Fig. 15. The construction is entirely brick/cement and incorporates no moving parts. It is possible to maintain an approximately constant gas pressure because in-

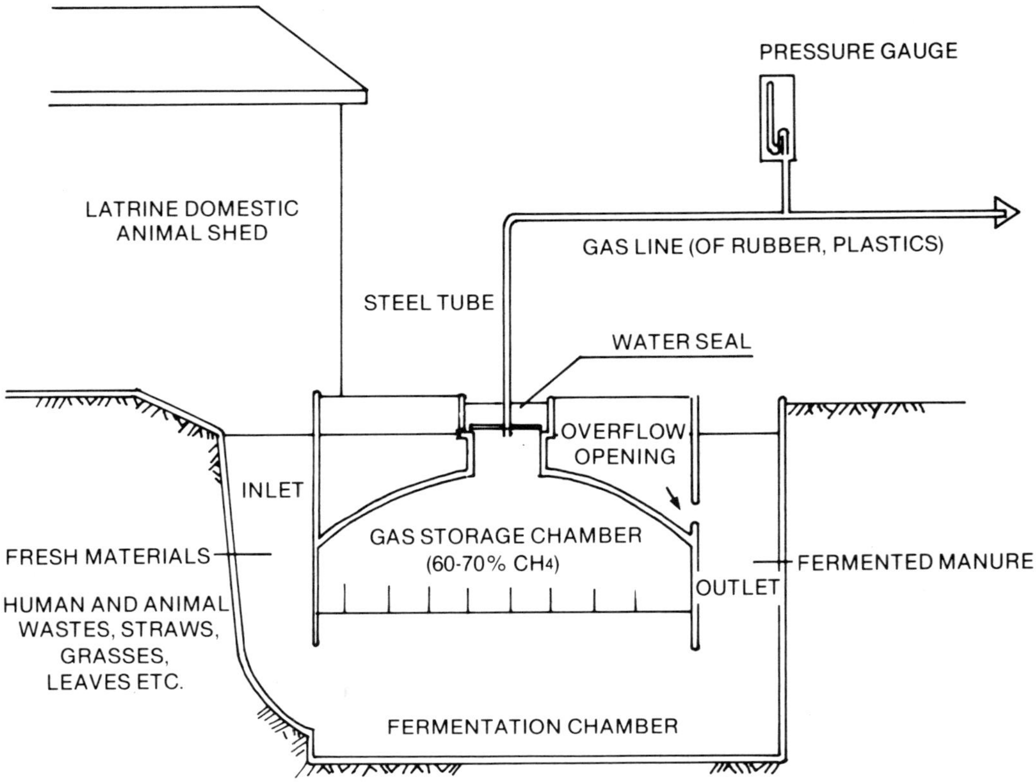

Fig. 15. This ingenious digester design of Chinese origin can handle animal manures and some vegetable wastes. It is constructed entirely of brick and cement and incorporates no moving parts.

creasing gas volume in the storage chamber expels some of the liquid content of the digester. If the cross-sectional area is large the change in liquid height and thus gas pressure is small. The only possible drawback lies in exposure of the fermentor contents to the air, but diffusion of oxygen into the digester slurry is usually negligible. The schematic diagram (Fig. 16) shows another variation: here water moves in response to changes in gas volume.

Neither cost nor operating data are as yet available for this digester. However, a range of designs covering a scale of operation from a single household to a whole community exist; thus, developments along these lines seem very promising.

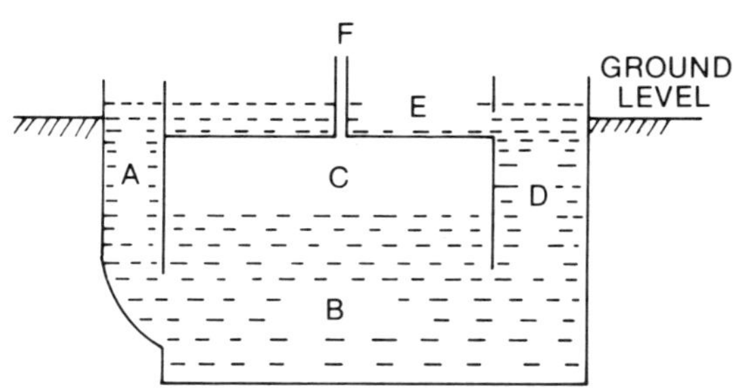

Fig. 16. General arrangement of an enclosed biogas plant in which the pressure of the biogas inside the gas storage tank is kept constant by automatic adjustment of the liquid pressure (A intake chamber; B fermentation tank; C gas storage tank; D outlet chamber; E liquid pressure tank; F gas pipe).

All the designs shown operate under a positive pressure: usually of a few inches water gauge; though in the case of the Chinese process the pressure in the gas can be several feet of water. One consequence is that the concentration of CO_2 dissolved in the fermentor (and thus pH) will be decreased. Whether this could depress the pH so far as to inhibit the methanogenic bacteria remains to be seen. Other authors have suggested that operation under a slight vacuum is beneficial (Sathianathan 1975), presumably because the dissolved CO_2 is lower and the pH higher. This is similar to the idea of Graef and Andrews (1974) to scrub CO_2 from the gas phase. Operation under vacuum is a hazardous business because of the possibility of air leaking into the digester, and it is not recommended. There are also (undocumented) reports that production is inhibited in deep-well digesters. This could be due to the increase in pressure at the bottom of a 5-m digester (the pressure would be about 1.5 atmospheres).

Sathianathan reports that maximum gas production is obtained (presumably in an unstirred digester) with diameter to depth ratios between 0.66 and 1.0. However, practical digesters often have ratios in the order of 0.25 (implications for heat loss from the digester must also be taken into account).

Effects of scale have not been properly studied. There are undoubtedly strong economies of scale as far as capital costs are concerned. There are also certain technical disadvantages in very small (3-4 m^3) plants. First, heat losses are high, which makes them uneconomical, and also makes it difficult to achieve stable temperatures. Small continuous plants also tend to be more unstable in operation because slight errors in feeding are magnified. It is also difficult to justify the expense of improvements in design (e.g. mixing, gas recirculation, etc., see Malina and Miholits), which could lead to substantially improved efficiencies. However, there may well be arguments for extremely small (oil-drum scale) units, to provide small gas outputs (enough to boil a gallon or two of water per day), since these can be constructed at close to zero cost, and may well be appropriate for individual families.

Alternative Treatment Systems Centred on Biogas

The alternatives within which anaerobic digestion could be the core are summarized in Fig. 17. The alternative treatments and end uses of the gas product from the digestion are summarized in Fig. 18.

It is convenient to separate the variations associated with handling, treating, and utilizing the liquid and solid wastes (Fig. 17) from variations associated with gas utilization.

The choice among alternatives depends on a number of system parameters, among which the crucial elements are the: quantities and types of waste available; forms of local social organization; objectives and priorities defined locally; regulations governing discharges, etc.; scale of operation; and opportunity costs of the inputs — fertilizer, fuel and power, land, water, labour, capital, etc.

Waste Treatment, Nutrient Recycling

An extended background discussion to this topic is given earlier. The alternatives sketched in Fig. 17 reflect, in approximate order of increasing technical complexity and integration, the ways in which the technical component may vary.

Alternatives 1 and 2

This is the case in which there is no processing. Attempts at improving these options must centre on: controlling nutrient losses by good farm practice, improving nutrient value to the land, and using in areas where contamination is negligible; and where evaporation and nutrient losses seem impossible to control, increasing burning appliance efficiency, and controlling hazard/contamination by e.g. oven design.

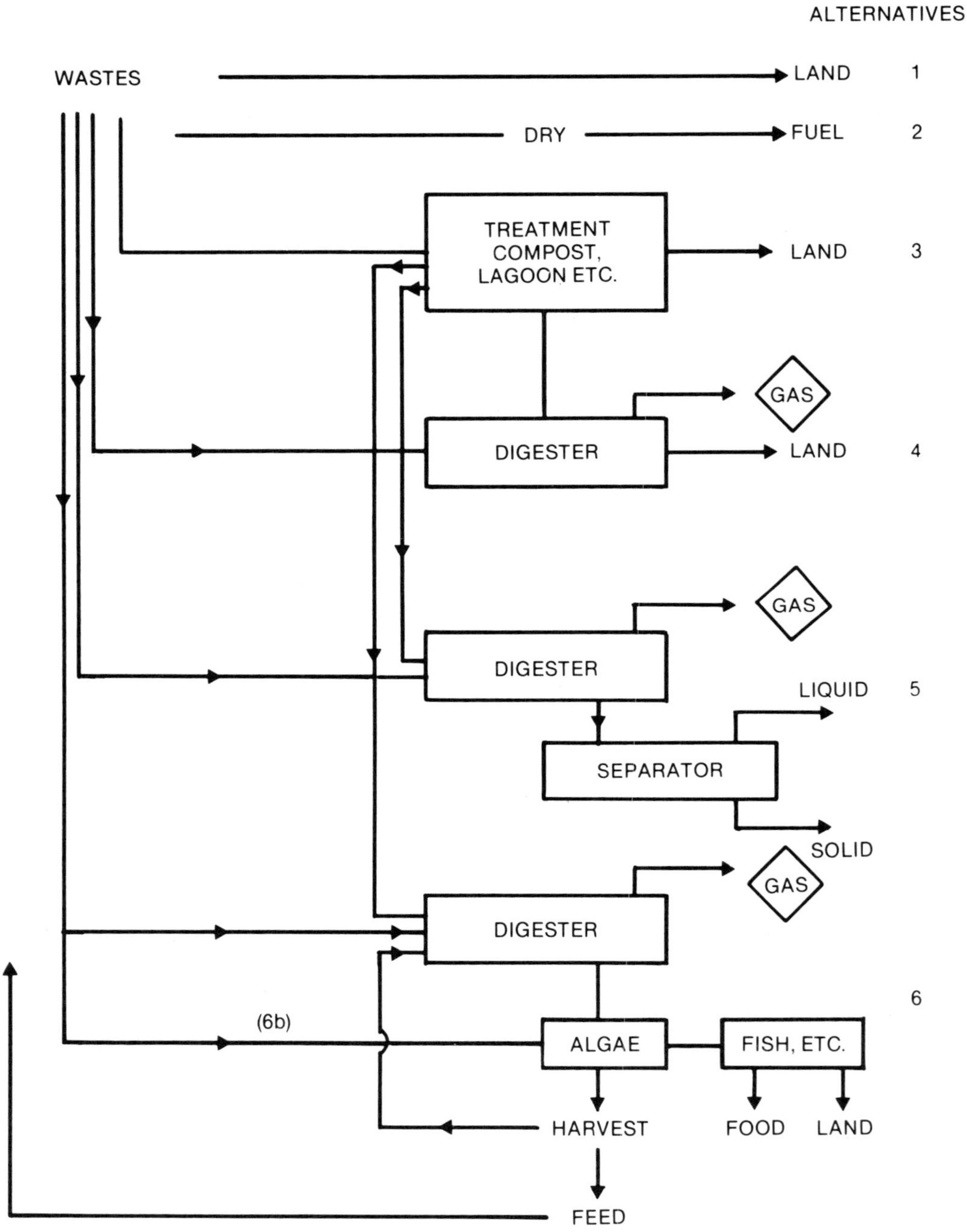

Fig. 17. Summary of alternative treatment systems within which anaerobic digestion could be the core.

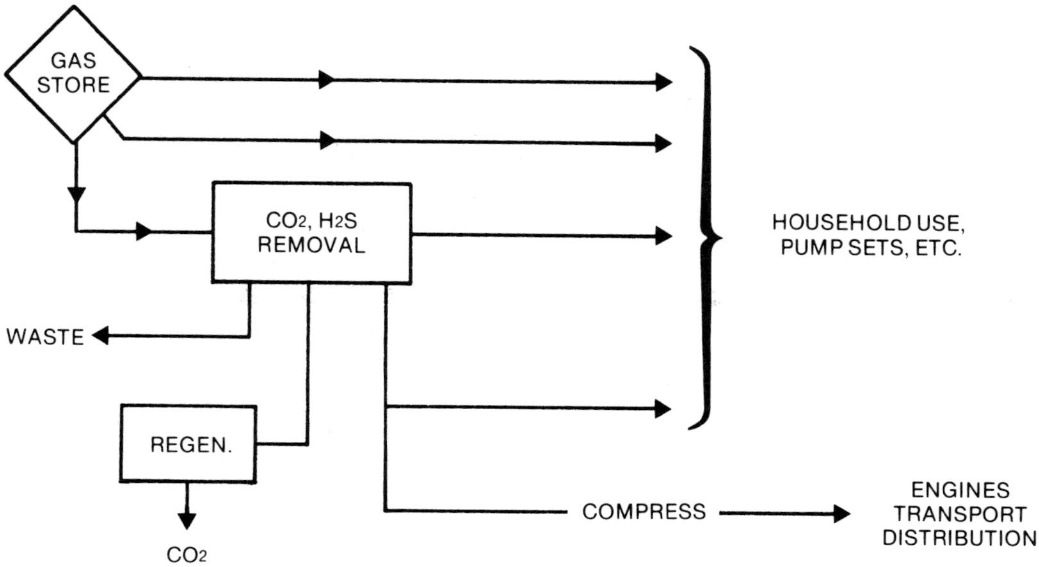

Fig. 18. Alternative treatments and end uses of the gas from the digestion process.

Alternative 3

The potential fuel value of the wastes is lost. This alternative can be regarded as an attempt to respond to the drawbacks of alternative 1 by improving the nutrient quality of the feed to the land (decreasing stream A in Fig. 19), and using the liquid waste more efficiently. The public health aspects are also improved.

Alternative 4

In its simplest form (waste → digester → slurry and gas) this is the 'core' technology that is the main subject of this report. Here the explicit attempt is to utilize both the potential nutrient and fuel functions of the waste (Fig. 20). There should be some trade-off between the fuel value A and nutrient value (stream B) of the products but this has not been seriously studied.

More refined core technologies (e.g. high-rate digesters) offer the benefits of higher efficiencies of conversion to gas and digested solid. So, too, does the use of pretreated waste as an input (i.e. inputs that have been macerated, partially composted, or decomposed). The net effect (at calculable cost) is to increase streams A and B in Fig. 20 at the expense of stream D. There still remains the question of the efficiency of utilization of the nutrient stream B, and this depends on the end use. Stream B could be used as a feed to a cereal crop.

Alternative 5

This improves the efficiency of the cycle and meets pollution control standards by making more efficient use of the liquid and solid streams; losses between treatment and cereal production are reduced.

Alternative 6

This attempts to make a more rational use of the waste stream. Here the only motivation for including a digester in the cycle is to obtain energy (as fuel) from the system, for sewage/animal wastes can themselves be used as direct feeds (variant 6b Fig. 17) (Shaw 1973; McGarry 1971; McGarry et al. 1972). In this system there are sufficient degrees of freedom to allow sets of objectives to be met more closely than in a simple once-through, one-unit process (alternatives 1 or 2). As Eusebio (1976)

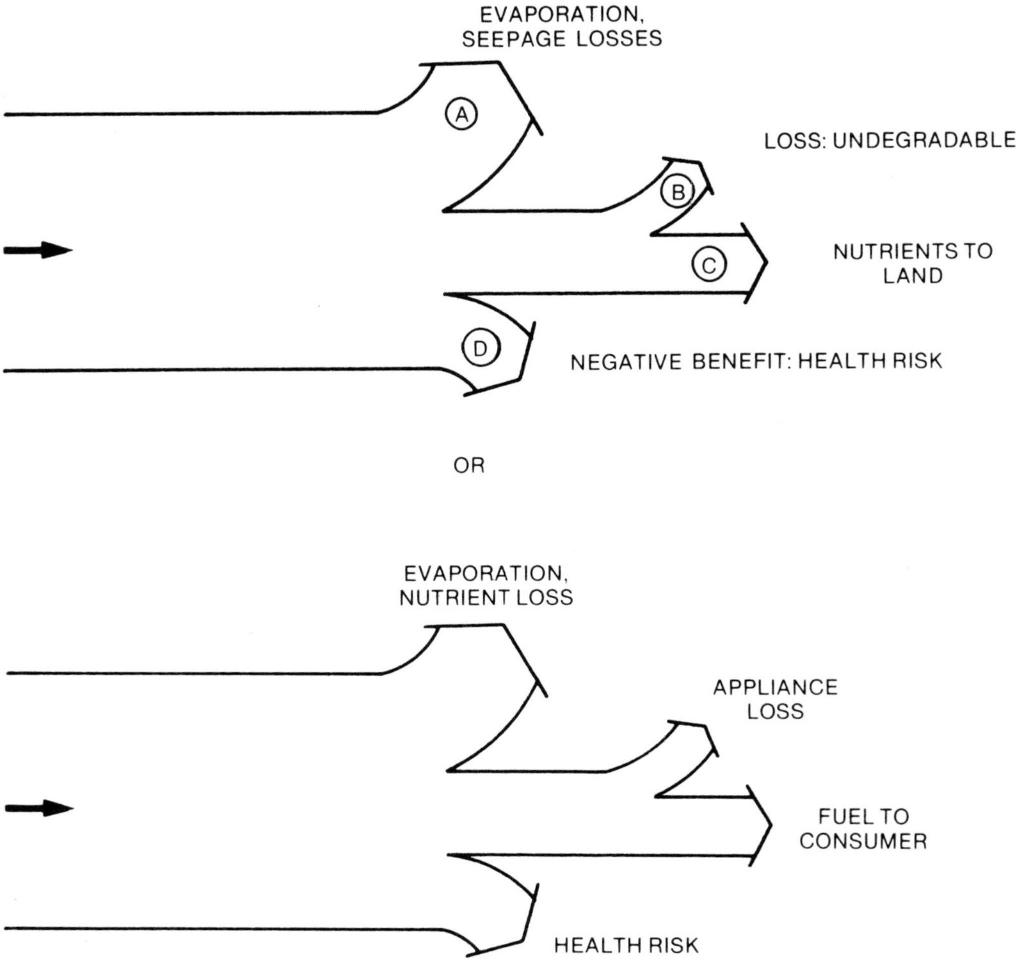

Fig. 19. Energy flows through alternatives 1, 2, and 3 of Fig. 17.

points out the basic ideas are: to meet pollution control requirements on the effluent; and to use the nutrients (especially organic N) more efficiently. Considering Fig. 20, for example, the two major losses or inefficiencies in the nutrient cycle are: loss due to seepage, evaporation, etc.; and inefficient take-up by the crop.

Alternative 6 attempts to minimize seepage loss by using the nitrogen in the liquid waste stream before it is returned to the land or discharged to a watercourse. In practice, and especially in the tropics, the use of algal ponds is one method of using to best advantage the local conditions, because high rates of algal growth are possible with photosynthesizing algae. The algae can then be used as a feed supplement for cattle or fish or fed to the digester. The nitrogen-containing water from the algal pond can also be used as a source of nutrients for fish or ducks (Chan 1973).

There is a danger of thinking that zero-cost solutions are possible or, alternatively, that integrated rural systems offer something for nothing. They do not. Every step in the processing stream has a degree of inefficiency; as one moves along the process stream so, inevitably, the marginal costs of recovering nutrients, for example, become more expensive. These processes must be evaluated carefully and realistically.

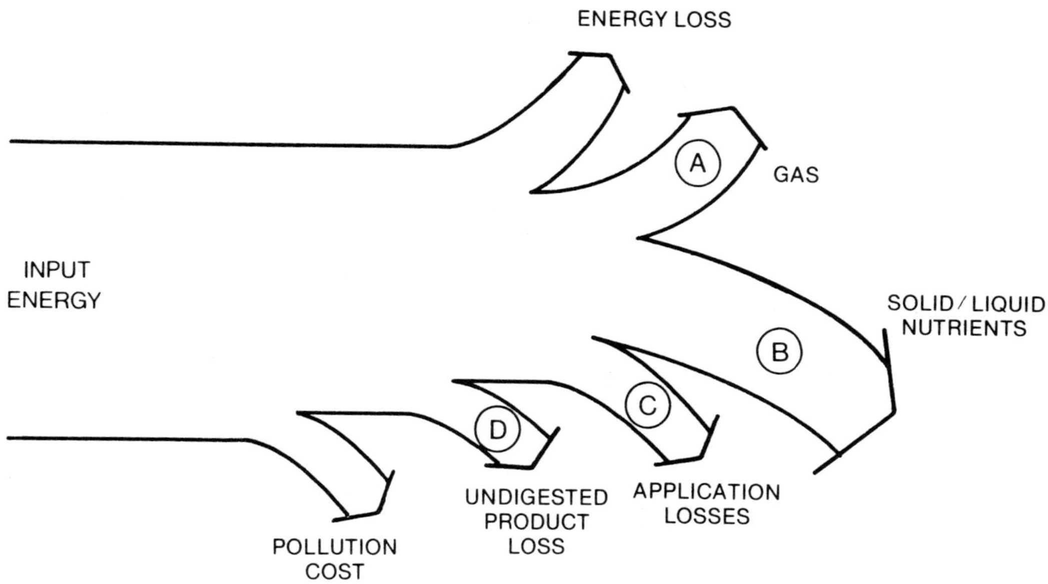

Fig. 20. Alternative 4 (Fig. 17) attempts to utilize both the potential nutrient and fuel functions of the waste.

Alternative Gas Handling/End Uses

Again, there is a scale of increasing complexity, sophistication, and cost (Fig. 18). The gas must be combustible (which implies rejecting gas produced early in the batch cycle or within a short time after start-up of a continuous tank). Under no condition should the gas be burnt until the process has settled down. The gas stream leaving the digester contains methane, carbon dioxide, negligible traces of other gases (H_2S, H_2), and is saturated with water. This gas cools along the pipeline and water condenses out in the line (just as it condenses on the inside of the gas holder). It is extremely important, then, that the pipes be at a slight angle to the horizontal and that provision be made for draining off the condensate.

The gas, even as a 50:50 methane/carbon dioxide mixture is combustible. Why then bother to purify the gas of CO_2? (The case for removing H_2S is strong because it is so noxious.) The arguments in favour of removing CO_2 for household use are that this will improve the burning properties of the gas (i.e. its calorific value and flame temperature). However, reliable burners for lean methane gases are available and there seems relatively little incentive for purification.

If the scale of production is sufficient (gas requirements can be calculated on the basis of a calorific value of the gas of 18-26 J/cm^3 see Tables 24 and 25) then various alternative end uses of the gas are possible.

Table 24. Comparison of the calorific value of biogas and other fuel gases (Meynell 1975).

	Calorific value (J/cm^3)
Coal gas	16.7-18.5
Biogas	20-26
Methane	33.2-39.6
Natural gas	38.9-81.4
Propane	81.4-96.2
Butane	107.3-125.8

NOTE: Variation depends upon degree of saturation and percentage composition of component gases.

Table 25. Volume of other fuels with a calorific value equivalent of 28 m^3 of biogas (at 22.2 J/cm^3 = 622 Mega J) (Meynell 1975).

	Volume of equivalent fuel
Natural gas	16 m^3
Liquid butane	24.3 litres
Gasoline	19.7 litres
Diesel oil	17.4 litres

The gas can be used to power internal combustion engines, pump sets, etc. (a good description of running efficiencies is to be found in Sathianathan 1976, p. 72f). The carbon dioxide acts as a diluent and affects the performance. Neyeloff and Gunkel (1975) point out the possible advantages of gaseous fuels: anti-knock qualities; absence of contamination and sludge formation; no need to add tetra-ethyl lead; and more homogeneous mixture conditions in the cylinder.

The same authors report the effects of CO_2:CH_4 mixtures on engine performance. The effects of increasing methane content on the specific power output are given in Fig. 21. Whether it is worth removing CO_2 depends on the trade-off between purification costs and improved performance.

If the engine is stationary (pump set, etc.) or used for local travel/power (e.g. a tractor) there is little incentive to compress the gas unless storage is a problem. If it is intended to use the methane as a fuel for a car then it certainly is imperative to minimize storage space and handling difficulties by compressing the fuel. Methane, unfortunately, does not liquefy easily (critical T and P: -82.5 °C,

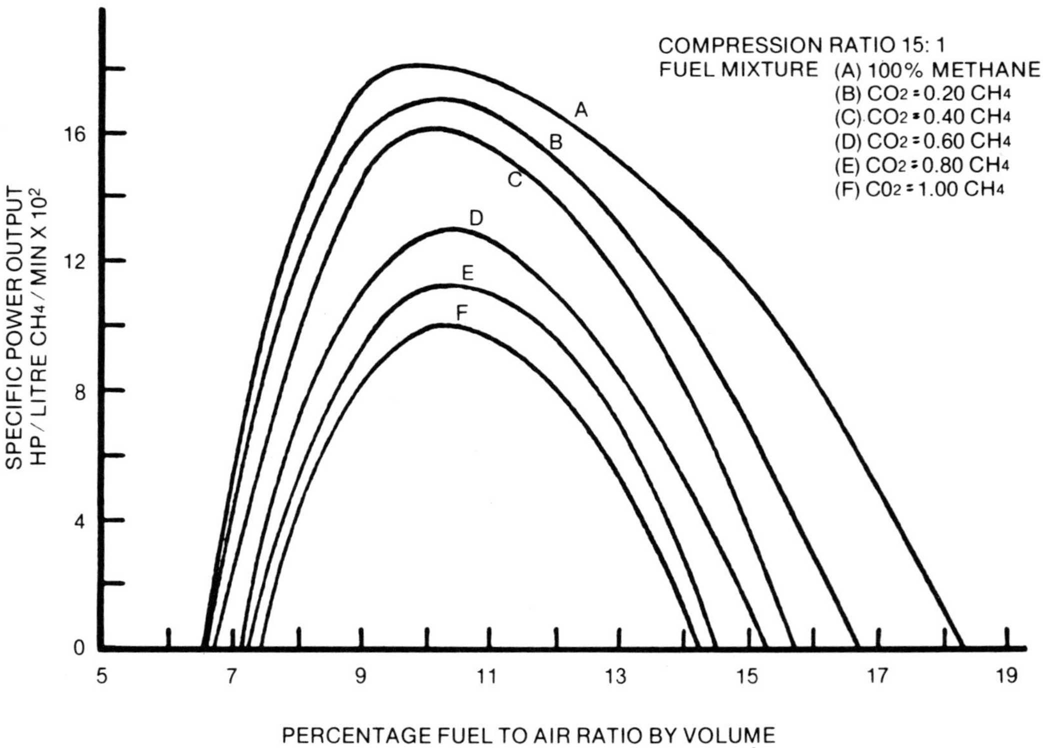

Fig. 21. Effect of various CO_2:CH_4 mixtures on engine performance as measured by specific power output (Neyeloff and Gunkel 1975).

46.0 bar) so that intermediate compression (perhaps using a simple single-stage compressor) to about 140 kg/cm^2 would be possible. Meynell (1975) showed that a cylinder 1.6 m × 0.27 m diameter would hold about 54 litres, weigh ~ 60 kg, and contain the equivalent of about 16 litres of petrol, i.e. approximately three times that needed for petrol storage. This assumes that the carbon dioxide has been scrubbed from the gas. If the objective of the biogas is to provide fuel for transport rather than to substitute for other fuels, one should instead consider alcohol production. Trevelyan (1975) gives a good description of the possibilities of alcohol as a fuel for combustion engines. Makhijani and Poole (1975) further discuss energy/fuel alternatives.

It is unrealistic to consider alcohol for household fuel, and there may be arguments in favour of using compressed bottled gas to serve a community from a central facility. The costs to evaluate these alternatives can be calculated on the basis of existing knowledge.

The possible reuse of carbon dioxide merits serious consideration. Carbon dioxide can be regenerated easily from lime water and could be used as dry ice for local health service, refrigerators, etc., or possibly to promote algal growth. These possibilities have not yet been evaluated seriously.

Technical Parameters Affecting Digester Performance

In this section the main variables and measures of digester behaviour are enumerated, and some attempt is made to summarize the state of knowledge on the significance of these variables. Later a number of the more important areas related to the assessment, design, and operation of digesters are reviewed.

Major Influences on Digester Performance

Many of these parameters and variables have already been introduced and discussed in the section dealing with alternative digester designs. We can separate the major influences into three broad groups: parameters characterizing the mode of operation; more specific design parameters; and inputs and possible disturbances. In view of earlier discussions, many of the variables are listed with little or no commentary.

Mode of Operation

This can either be a batch (mixed, partially mixed, or unmixed) or continuous (mixed, partially mixed, unmixed, plug flow, or anaerobic contact) operation.

Design Parameters (Associated with Fabrication etc.)

Materials of construction; configuration, length/diameter ratio; number of stages; and heating arrangements are involved.

Inputs

Processes (especially continuous ones) depend on various inputs and are subject to intermittent disturbances due to fluctuations in environmental conditions (temperature, rainfall, etc.) or in feed materials (composition, quantity, operator errors, etc.). It is extremely difficult to monitor or control some of these inputs, and it is important to know the relative importance of the main inputs and the sensitivity of the process. It is also necessary to devise methods to monitor process performance, detect incipient malfunctions, and to correct malfunctions (see Table 26).

The acceptable ranges of the majority of these variables have been discussed in earlier sections. Others are discussed in more detail below.

State Variables

The most significant measurable variables that reflect (and influence) digester behaviour are given in Table 27, along with an indication of the measurement technique needed.

In practice, only a few of the state variables are measurable on a day-to-day

Table 26. Inputs and disturbances of digestion process.

Parameter	Easily measurable?	Possible control variable
BOD, COD feed	No	No
Feed composition	Difficult	Yes[1]
Feed: physical state, size	No	No
Feed concentration (solid:liquid)	Yes: approximately	Yes[2]
Retention time	Yes	Yes[2]
Loading rate	Possible[3]	Yes[2]
Bacterial, or seed content	No	Yes[4]
Feed temperature	Yes	Yes[5]
Toxic materials	No	No
Nutrient content	No	Yes[6]
C/N ratio	No (but calculable)	No
Heat input	Indirectly	Yes
Heat losses	No	No
Pressure	Yes	?[7]
Ambient temperature/conditions	Impractical	No
'Secondary' disturbances[8]		
Gas composition		?
Digester temperature		—

NOTE: A rather subjective opinion has been taken as to what is 'easily' or 'cheaply' measurable. The categories are thus *extremely* subjective and inexact.

[1]By adding lime, urea etc. to control pH or stimulate the operation.

[2]Little is known of the dynamic effects of (small) changes in these variables, which are, of course, interrelated.

[3]Dry solids could probably be measured relatively frequently but monitoring of loading rate is impractical in rural application.

[4]By using 'starters' or recycles — again, there is little experience to draw on.

[5]At some cost.

[6]By adding known quantities of specified nutrients.

[7]Effect as yet unexplored.

[8]These are really 'state' (i.e. dependent variables) which themselves can affect further behaviour.

basis. Clearly, the operator will recognize the symptoms of a failing or malfunctioning digester — usually, in the first instance, through a fall in gas production. Some variables are less important or sensitive than others. Given the complex microbiology of the fermentation process it is highly interacting in the sense that few if any input or control variables affect only one measured or state variable. At different feed compositions, the acclimatized bacterial population will presumably be different, leading to different alkalinity and pH conditions and ultimately to different gas compositions and production rates. Similarly, a change in feed composition will trigger changes in all these variables. The more significant state variables or 'indicators' are discussed below.

State-of-the-Art Review

The important areas relevant to digester design, operation, and use that are outlined in this section are: technical feasibility data; the microbiology/bacteriology of anaerobic fermentation; the kinetics of digestion; engineering design aspects; operation and control of digesters; gas handling and use; instrumentation for operation and control; and problems related to 'peripheral' technologies — oxidation ponds, etc.

Feasibility Data

If a given substrate or substrate mixture is biodegradable to methane under anaerobic conditions, there are a series of supplementary questions to be asked: Under what conditions does the process work (best)? What is the likely methane yield per unit weight of substrate? What are the water, energy, and nutrient requirements? What is the slurry production rate (and its characteristics)? What will be the measured operating conditions under normal operation? What size of equipment is needed? It would be advantageous to be able to answer all these questions to within a defined degree of accuracy when assessing the feasibility of a proposed project.

There is enough information on digester operation and sufficiently well-proven basic information to be able to answer most of these questions.

Biodegradability of Substrate

There is no completely adequate theory of biodegradability. Nonetheless there is a con-

Table 27. The most significant measurable variables that reflect (and influence) digester behaviour.

Variable	Easily (cheaply) measurable?	Recommended acceptable range	Measurement technique
Temperature	Yes	10-60 °C	Thermometer etc.
pH	?	6.4-7.5 (→8.0)	pH meter; litmus
E_h	No	<-330 mV redox pot.	Instrument
Alkalinity	?	2000-35000 mg/litre ($CaCO_3$)	Titration
Toxic materials	No	depends on species	Specific technique
Gas production rate	?	—	Meter
Gas composition	?	>50% CH_4	Orsat analysis
Nutrient levels	No	depends on nutrient	—
BOD, COD	No (?)	(Variable + depends on water regulation)	Laborious analysis
Cell content	No	— (higher the better?)	—
Bacterial population	No	—	—

NOTE: A rather subjective opinion has been taken as to what is 'easily' or 'cheaply' measured. The categories are thus *extremely* subjective and inexact.

The variables noted with a question mark are ones that can be measured relatively easily, but may not yet be feasible within the context of village technology.

siderable amount of laboratory and practical information on the anaerobic decomposition of organic materials. Lignins are degraded slowly if at all; insoluble compounds degrade more slowly than do soluble ones. In practice, animal and human wastes can be degraded and many vegetable and crop residues and wastes from agricultural processing can be fermented. Good sources of information are Meynell (1976), Sathianathan (1976), Buswell and Boruff (1932), and the series of papers from ICAR. Mosey (1974) covers the processing of industrial and urban wastes.

Conditions

The major indicators or determinants are the concentration and composition of the feed, the temperature, and the pH. There is little variation in these parameters among different feedstuffs, and concentrations of up to 10% (dry weight) can usually be handled. It may be necessary to adjust the composition of the feed to ensure that the process is not limited by the lack of a particular nutrient. The simplest overall measure of chemical composition is the carbon/nitrogen ratio, which gives some guidance as to the range of feedstuffs that can be handled. Generally C/N ratios in the range 10-30 are recommended, but this figure is not absolute. Typical C/N ratios of some feeds are given in Table 28.

Table 28. Typical carbon/nitrogen ratios of some feeds.

	N (% dry weight)	C/N ratio
Night soil	6	6-10
Cow manure	1.7	18
Chicken manure	6.3	7.3
Horse manure	2.3	25
Hay, grass	4	12
Hay, alfalfa	2.8	17
Seaweed	1.9	79
Oat straw	1.1	48
Wheat straw	0.5	150
Bagasse	0.3	150
Sawdust	0.1	200-500

Temperatures in the range 15-60 °C can be used, but the temperature is usually chosen to be 30-35 °C (see earlier comments).

The equilibrium pH, which should be approximately 7-8 (i.e. slightly alkaline), will be established in a self-regulating way when the process functions correctly (see also "Equipment size").

Likely Yields

Yields depend on the detention or batch time. Typical detention times for continuous processes in conventional digesters are approximately 30 days, when operation efficiencies (expressed as percentage destruction over the digester) are about 50-70%. Under these conditions some typical gas yields are presented in Table 29.

Table 29. Typical reported yields from anaerobic digesters.

	Gas yield (m^3/kg volatile matter fed)	Gas composition (% methane)
Cow dung	0.09-0.3	65
Chicken manure	0.3	60
Pig manure	0.35-0.48	65-70
Farm wastes	0.3-0.42	60-70
Elephant grass	0.42-0.54	60
Chicken manure/ paper pulp	0.42-0.48	60
Chicken manure/ grass clippings	0.35	68
Sewage sludge	0.6	68

Appendix 1 gives a comprehensive listing of experimental data, and Table 20 operating data.

Knowing the quantity of feed available and its approximate composition gas yields can be calculated. It is interesting to see how far the data can be based on fundamental principles.

Limits on Performance

The efficiency of a biogas plant depends on many factors — the design, operating conditions, raw materials, etc. This variability is shown in the published data and has occasionally led to outrageous claims for potential yields. It would be useful to be able to set bounds on performance to assist in feasibility studies, check claimed behaviour, etc.

Maximum Gas Yield

Buswell and Mueller (1952) produced a simplified overall picture of the anaerobic fermentation of a typical substrate ($C_nH_aO_b$) to carbon dioxide and methane. The overall stoichiometry is oversimple (for example, it neglects cell formation), but it represents the limit of what could happen. Their equation is

$$C_nH_aO_b + (H_2O) \rightarrow$$

$$\left(\frac{n}{2} - \frac{a}{8} + \frac{b}{4}\right) CO_2 + \left(\frac{n}{2} + \frac{a}{8} - \frac{b}{4}\right) CH_4$$

The composition of the gas depends on the substrate, and, in principle, is predictable. The total gas yield ($CO_2 + CH_4$) can also be calculated, a priori, because 1 kg of carbon in the substrate will yield 1/12 kmole gas product. Thus, per kilogram of carbon decomposed, the yield of gas should be (22.4/12)m^3 gas (measured at STP) or 1.867 m^3 gas.

On this basis, Table 30 has been derived, using average carbon contents of the materials. The values can be interpreted in two ways: first, as the maximum gas yields possible per unit mass dry matter fed to the digester; second, as the maximum gas yields possible per unit quantity of dry matter destroyed. It is assumed that none of the substrate leaves as cellular matter or intermediate volatile acids. It is also assumed that all the carbon in the feed is susceptible to anaerobic digestion. Actual yields (Table 29) are, of course, somewhat lower than the calculated values. The tabulated values, however, give a rough guide to the effect of substrate on yield.

Considering a typical continuous digester, with feed containing around 2% volatile solids (of which C = 50%), the carbon content in the feed will be approxi-

Table 30. Maximum gas yields (from various sources).

	N	C/N	C%	Gas yield (ft^3/lb DM)
Feces	—	—	40-55	22.4-30.9
Blood	10-14	3	30	16.8-23.5
Young grass clippings	4	12	48	26.9
Lucerne	2.4-3	16-20	<60	<33.5
Grass clippings	2.4	19	45.6	26.5
Manure (large)	2.15	14	30.1	16.86
Seaweed	1.9	79	36.1	20.2
Oat straw	1.05	48	50.4	28.2
Wheat straw	0.3	138	38.4	21.5
Sawdust	0.11	511	56.2	31.5
Carbohydrate	—	—	—	12.0
Fat	—	—	—	23.1
Protein	—	—	—	15.7
Horse manure	2.3	25	57.5	32.2
Cow manure	1.7	18	30.1	17.22
Hay	4	12	48	26.9
Pig manure	3.8	20	76.0	42.5
Sheep manure	3.8	22	83.6	46.7
Poultry	6.5	15	~90	50.3
Garbage	3	—	54.7	30.5
Paper	—	—	40.6	22.8
Newspaper	0.05	—	40.6	22.8
Chicken manure	3.2	—	23.4	13.2
Steer manure	1.35	—	34.1	19.1

NOTE: To convert ft^3/lb to m^3/kg multiply by 0.62.

mately 1%, which is approximately 1 kmole/m^3 feed. With a loading rate of X kg VS/m^3/day the maximum gas yield would thus be 1.867 ($X/2$) digester volumes/day. or, very approximately, X digester volumes/day. Loading rates vary from 0.8 to 3.5 kg VS/m^3/day, so the maximum gas yields that can conceivably be produced range from 0.8 to 3.5 digester volumes/day. Thus, there is considerable economic incentive to seek designs that operate at high loading rates.

It is possible that some of the CO_2 and CH_4 may be lost in the slurry. For a 50:50 gas mixture at 1 bar at 35 °C, there will be 0.2 litre dissolved CO_2 and 0.01 litre CH_4 (measured at 0 °C, 1 bar) per litre slurry. Thus, in a plant operating at a loading rate X kg/m^3/day and producing X digester volumes gas/day, the effluent slurry will carry with it some (0.01/8) digester volumes of CH_4 per day, which is negligible in comparison with the total gas production rate.

Gas Yields and COD Removal

Very often the quality of an effluent is measured in terms of its COD value. Using the Buswell and Mueller stoichiometry, if the substrate were oxidized completely it would require (n + a/4 - b) kmole oxygen per kmole substrate,

$$C_nH_aO_b + \left(n + \frac{a}{4} - \frac{b}{2}\right) O_2 \rightarrow nCO_2 + \frac{a}{2} H_2O$$

Moreover, one kmole substrate digested anaerobically should yield (n/2 + a/8 - b/4) kmole methane, so that 1 m^3 methane produced is equivalent to

$$\frac{1}{22.4}\left[\frac{(n + a/4 - b/2)}{(n/2 + a/8 - b/4)}\right] \times 32 \text{ kg } O_2 =$$

$$2.85 \text{ kg COD}$$

In other words, it is possible to calculate the equivalence between COD removal and methane generation directly from first principles. However, because some carbon must go to producing cells and volatile acids this is likely to underestimate the COD reduction.

For example, Chung Po et al. (1974) found that the process efficiency varied from 1.2 to 1.4 m^3 gas per kg COD (pig swine digester), which, assuming 50% methane, is equivalent to 1.7 kg COD/m^3 methane.

Water, Energy, Nutrient Requirements

Water requirements depend on the concentration of the input stream and the hydraulic retention time. Typical Indian designs operate with a water:animal manure ratio of about 1:1, which is equivalent to a dry matter concentration in the feed of about 9%. Other designs (e.g. Richard 1975; Chan 1973) operate with dilute feeds and very short residence times (a few days) so that water consumption rates are extremely high. If water is relatively expensive or scarce attempts should be made to economize or reuse it.

The energy required to maintain stable operation at a desired temperature can be calculated with some confidence from basic principles. The principle sources of heat loss from the system have been discussed before. Each term can be calculated directly provided good estimates of local environmental and subsoil conditions (especially temperature) are known (examples will be found in Jewell 1975). Other thermodynamic properties are discussed below.

Calorific Values of Substrate and Product, and Heat of Reaction

McCarty (1964) discusses the thermodynamic implications of the anaerobic digestion process. Again, a simplified approach yields useful information about the overall feasibility of the process.

$$C_nH_aO_b\ (+ H_2O) \rightarrow \left(\frac{n}{2} - \frac{a}{8} + \frac{b}{4}\right) CO_2 + \left(\frac{n}{2} + \frac{a}{8} - \frac{b}{4}\right) CH_4$$

If the standard heat of combustion of the substrate is $(-\Delta H_{cs})$ then, because the standard heat of combustion of methane is -88 345 kJ/kmol, the standard heat of reaction of the fermentation is

$$(-\Delta H_{cs}) + (n/2 + a/8 - b/4)\ 88\ 345 \text{ kJ/kmole substrate}$$

or

$$\frac{(-\Delta H_{cs})}{(n/2 + a/8 - b/4)} + 88\ 345 \text{ kJ/kmole methane}$$

(neglecting heats of solution).

In general, the process is mildly exothermic. The higher the proportion of substrate diverted to carbon dioxide, the higher the heat release but this is of little avail in practical circumstances.

On the basis of these figures, one can calculate heating and insulation requirements. Such calculations can be carried out from first principles.

In practice, it is usually assumed that with organic wastes of natural origin there is no need to provide additional nutrients (if the C/N ratio is approximately correct). This assumption will not be true in the case of many industrial wastes.

Slurry Production Rate

This will be very nearly equal to the volumetric flow rate of input matter. The slurry contains all the nitrogen from the input, and some 50% of the input organic matter.

Measured Operating Conditions

If designed and operated properly, the temperature should be close to the desired value. Gas flow rate (and composition) have been discussed above. The pH should settle down to a value around 7.2. Trevelyan (1975) shows that it is possible to calculate the pH, using basic information on the buffering capacity of the process. For example, if protein were fermented at a concentration of 1 g atom C/litre, the final pH would be above 8, and would lead to problems of ammonia toxicity. The possibilities of toxicity can also be calculated from first principles (Mosey 1974).

In other words, assuming complete decomposition of the substrate to methane, carbon dioxide, and water, useful limits can be put on the digester performance.

Equipment Size

This cannot be calculated from first principles in any rigorous way for the substrates of interest because this requires a knowledge of the process kinetics. In practice, it can be taken that a simple unit will have a gas production rate of about 0.5-1 digester volumes per day. Knowing the gas yield per unit quantity of substrate it is possible to put an approximate size on the equipment, but there is a good deal of room for improvement in this respect.

Microbiology

There are a number of excellent up-to-date reviews of the microbiology of anaerobic fermentation (e.g. Hobson et al. 1974; Trevelyan 1975). Rather than present a detailed review here, the main features of the microbiology as presently understood are discussed, with special reference to the implications for digester behaviour, future developments, etc.

Anaerobic microbial metabolisms may take place whenever the supply of oxygen is stopped or is so limited that aerobic processes quickly remove the oxygen. Thus, it takes place below the surface in still waters, ponds, or lagoons. Generally it is characterized by extremely small energy changes per unit substrate decomposed (McCarty 1971). Moreover, the overall process can be approximately divided into three sequential stages, of which the first two are so intimately linked that they are often considered together. Thus, the relatively small energy yield from the overall conversion is divided into even smaller packets and distributed among the different bacteria involved. As a result, the production of solids (i.e. microbial cells) is small, which is particularly advantageous for waste stabilization and disposal (Pfeffer 1966). In aerobic processes, on the other hand, energy changes are large and the quantities of microbial solid to be disposed of are often embarrassingly large.

In practice, most organic wastes consist of a range of materials (carbohydrates, proteins, lipids, fats, and salts), and the general scheme (adapted from Hobson et al. 1974) of their fermentation is shown in Fig. 22.

Ideas based on the behaviour of pure bacterial cultures in the presence of single pure substrates are likely to have rather limited application because a balanced microbial flora, dependent on the feed, seed, etc., is essential to the process. One of the factors determining the composition of the mixed culture flora (some constituents of which are essential while others are present fortuitously) is the energy available from the biochemical reactions that are spread among the bacteria. A mixed culture acts synergistically, that is it can do more than is estimated by 'summing' the effects of pure cultures acting on single substrates. For example, the presence of additional substrates or bacterial strains can modify the process yield (by suppressing or accelerating the degradation of a particular substrate) (Hobson et al. 1974, p. 147). An essential feature of the behaviour of mixed cultures, which is so far incompletely understood or studied, lies in the interactions and interdependence among different bacterial strains.

There are also difficulties in transferring results from one set of anaerobic conditions to another. For example, the multicomponent stomach of ruminant animals is among the oldest established anaerobic processes. Optimum conditions in the rumen are such as to minimize methane production and in practice the rumen has developed to

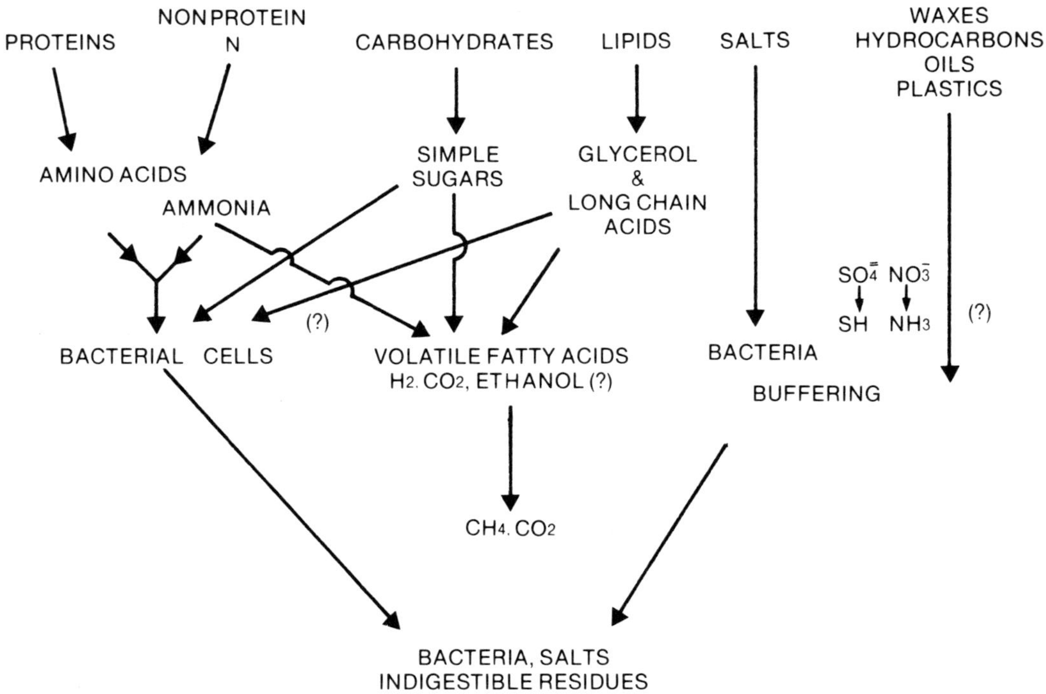

Fig. 22. Most organic wastes consist of a range of materials that can be fermented following this general scheme (adapted from Hobson et al. 1974).

that end (nonetheless, Trevelyan quotes typical methane yields from a cow of 100-500 litres/day, or 5-10% of the calorific value of the diet!). On the other hand, optimum design of a digester seeks maximum methane production and the minimum production of acids and microbial cells.

The first stage of digestion is the hydrolysis (by extracellular enzymes) of complex organic substances to soluble monomeric or dimeric compounds (e.g. cellulose, glucose). A wide range of cellulolytic and other bacteria have been identified and related to this stage; their population depends on the feed composition. The cellulolytic bacteria are often classified in two groups: the mesophilic bacteria, which have an optimum temperature range of about 35-40 °C, and the thermophilic, with an optimum about 55-60 °C. Another important feature is that the synergistic (or cooperative) action of these bacteria can lead to a faster removal of cellulose than by pure cultures. The optimum pH range for the bacteria is in the range pH 5-7.

There is good evidence (for example, Hobson et al. 1974; Chan 1971; and recent work in our own laboratories, Hadjitofi 1976) that cellulose hydrolysis is often the slowest (rate limiting) step in anaerobic digestion. Hobson also reviews the likely processes in the breakdown of proteins and lipids. The simpler compounds resulting from this first stage of digestion serve two functions: they contribute to the overall reduction and stabilization of the waste, and they are vital sources of energy and cell components for the bacteria.

A good deal is known of the main requirements for bacterial growth and function: that is energy (via organic compounds), nitrogen, and various trace elements and salts. The popular literature may be rather misleading in this respect because a good deal tends to be made of the nitrogen requirements (as expressed in C/N ratio). This is not as crucial as is sometimes suggested, and usually only with rather specialized industrial wastes are these major components seriously out of balance. There has been little reported work on the nutrient requirements or supplementation necessary for wastes typical of developing countries.

In stage two the carbohydrates resulting from the first stage are fermented to one or more of: hydrogen; carbon dioxide; formic, acetic, propionic, butyric, valeric, lactic, and other acids; and simple alcohols. This stage is the principal source of energy for the bacteria in the digester; however, the microbiology of the processes involving acid-forming bacteria is incompletely understood. The proportion of the different products from stage two depends on the flora present, the substrate composition, and the environmental conditions. These will depend to a considerable extent on the rate of hydrolysis (i.e. stage one). Whether all, or most, of the intermediate products listed above can be attacked directly by methanogenic bacteria is still a contentious point. Some authors consider that the only substrates for the final stage are carbon dioxide, hydrogen, formic acid, and acetic acids. Acetate (a salt or ester of acetic acid) is often the single most important intermediate (Smith and Mah 1966 state that 73% of the methane originated from acetate in the digestion of sewage sludge). It may be, as Trevelyan (1975) notes, that the reactions of other acids are coupled together, so that the overall picture is of several methane producers acting serially.

In the third, or methanogenic stage the soluble products are converted. The energy involved in these reactions is small and in consequence the amount of bacterial cell formation is also small; on the other hand, some of the ammonia in the liquid resulting from stages one and two is utilized by the methanogenic bacteria. In fact, the methanogenic bacteria are completely dependent on the primary stage bacteria for growth. Besides depending on them for the provision of nitrogen (as ammonia) and the limited number of substrates that can be utilized, an oxidation-reduction potential (E_h) below -330 mV is needed for growth. In mixed cultures, the metabolic activities of the facultative anaerobes in the primary stages serve to reduce the E_h to the required level; the methanogenic bacteria themselves cannot produce these reduced conditions.

Only a very few methanogenic bacteria have been isolated in pure culture. The first strain considered to be pure (*Methanobacterium omelianski*) was subsequently shown to be a symbiotic association of two species, one producing acetate and hydrogen from ethanol and the other using the hydrogen to reduce carbon dioxide to methane. The five known pure strains all reduce carbon dioxide by hydrogen to methane

$$CO_2 + 4H_2 \rightarrow CH_4 + 2H_2O$$

Four can convert formate to methane, but probably via hydrogen and carbon dioxide. These two simple reactions are held to be responsible for all the methane production in ruminants (where the turnover time is about 1 day). This observation can be reconciled with earlier remarks about the importance of acetate as an intermediate because the growth rate of the acetate-utilizing bacteria is much slower than that of the hydrogen-utilizing bacteria. In a digester with a residence time of a day or two, acetate-users are bound to be unimportant; at retention times of, say, 20 days the picture may change considerably, as the bacterial population itself changes. Thus one must beware of over-simple statements that the methanogenic bacteria have slow growth rates (which ones?). As Hobson et al. (1974) point out, the question of growth rates in natural habitats is extremely complex.

Slow-growing bacteria are often very sensitive to shock changes in operating conditions, which may well lead to digester failure. Indeed, digesters are generally less able to cope with rapid changes in temperature, feed composition, acidity, etc., than with slow changes. Presumably this is related to the doubling time of the bacteria, which in the case of some methanogenic strains is of the order of several days. This suggests that the system will encounter severe difficulties in adapting to changes on a much shorter time scale than this. It is this sensitivity, usually manifested in falling pH as acids accumulate that gives rise to a 'stuck' or 'slow' digester, which is a major problem in continuous fermentation. (Another sign of incipient failure is either (or both) a falling gas production rate and/or an increasing carbon dioxide concentration.)

The methane bacteria are extremely sensitive to some factors. They are obligate anaerobes and their growth is inhibited by small amounts of oxygen or an oxidizing agent. They are slow growing (it has been generally accepted that growth rates from 4 to 10 days are typical, although Ghosh and Pohland (1974) argue convincingly that generation times are of the order of 5 hours), and are at a disadvantage because of their low numbers. They are particularly sensitive to pH. Methane production is satisfactory between pH 6.6 and 7.6 (Dague 1968), but methane formation is inhibited below 6.6 and conditions become toxic below 6.2. The first group of bacteria will continue functioning until pH 4.5, and Borchardt (1971) found that with great care methane formation continued down to such levels. This can be considered an exception, however.

The pH of the system depends on the rate at which intermediates are fermented to methane and carbon dioxide, i.e. on the alkalinity and volatile acid concentration. It probably makes little sense to talk of an optimum pH because this is the integral result of the different contributions from the various reactions; moreover, the optimum pH levels for the separate stages of the processes could be different.

The system can usually 'absorb' fluctuations in acid or base concentrations because of the natural buffering provided by the ammonia and bicarbonate ions. The buffering provided by the carbon dioxide/bicarbonate system is represented by

$$\text{pH} = 6.3 + \log (HCO_3^-)/\text{dissolved } CO_2)$$

The concentration of dissolved carbon dioxide depends on the temperature and partial pressure (pCO_2) (i.e. volume fraction of CO_2 in gas above the fermentor × total pressure). Typically, at 35 °C the concentration of dissolved CO_2 = 0.592 pCO_2 litres/litre water. Thus, gas composition and operating pressure affect the pH and, ultimately, digester performance. If the acidity in the digester begins to build up (i.e. pH falls), the proportion of CO_2 in the gas increases, leading to a further drop in pH. In other words, the system has a limited degree of self-regulation, and it is easy to see how the system becomes unstable.

It is advisable to maintain a moderate total alkalinity (as $CaCO_3$) (values of 2000-35000 mg/litre are usually suggested); at low values, a slight increase in volatile acid concentration leads to a large drop in pH. On the other hand, at high values the ammonium ion dissociates (to NH_3 and H^+). A number of authors (McCarty 1964; Mosey 1974) consider that toxicity may be due to free ammonia, and Mosey has reported the conditions for toxicity (as a function of pH), with an upper limit of 3000 mg/litre N as NH_3. If the process is thermophilic and the substrate contains a high proportion of protein, the system could be self-toxic; Trevelyan (1975) illustrates this and shows the value of simple overall calculations in predicting such conditions. Other forms of toxicity (e.g. due to the presence of salts or heavy metals) have been considered and it may be concluded that reasonable guidelines exist to help the unwary (Mosey 1974). One important consideration relates to attempts to control the pH. If the pH falls, it is often suggested that lime be added; however, lime reacts with CO_2 to produce calcium carbonate and at alkalinities above about 1000 mg/litre this produces an insoluble deposit. McCarty (1964) argues that sodium bicarbonate is a far better buffer. As Mosey (1974) notes, lime has the dual disadvantage of removing an important substrate for the bacteria (CO_2) and increasing the likelihood of scale formation in the digester.

In the event that the pH needs to be decreased, hydrochloric acid can be used (but not sulfuric or nitric acids).

Kill Rates of Pathogens

The main vectors and causative organisms in fecal-borne diseases were summarized earlier. Some of the available results on the kill rates of pathogens during anaerobic fermentation are summarized in Table 31. With digester temperatures above 35 °C and detention times of greater than 14 days most vectors will be destroyed. The eggs of the roundworm *Ascaris lumbricoides* are a major exception.

In Europe, sludges are occasionally pasteurized before discharge. This appears infeasible in most developing countries. Under normal operating conditions the public health control aspects of anaerobic digesters handling human excreta are comparable with any other feasible technique. However, the behaviour of digesters operating at very low retention times should be studied carefully.

Likely Developments

A major characteristic of anaerobic digestion in practice is that the process depends on an acclimatized mixed culture of bacteria. Very little is known of the population dynamics or ecology of these cultures (see Hattingh and Toerien 1969), and still less of ways in which particular strains might be encouraged or suppressed if this were useful in improving process efficiency. The room for improvement in speeding up the fermentation process and for improving its robustness or stability is enormous and it may well be that unless such improvements can be made biogas fermentation will always be at best a marginal contributor to rural and industrial development. Nutrient requirements and the ways in which different species compete for limited nutrients are little understood; at a practical level there is little information on nutrient requirements and the returns on them.

Doubtless, as the role of the different bacterial strains becomes clearer, it will become possible to devise methods or to create environments to give improvements in efficiency. This will require a good deal of microbiological and empirical work of the highest order.

The Rate of Methane Generation

As discussed earlier, the rate of methane production (a major determinant of the digester volume) depends on a wide range of parameters. Ideally, one would like to have simple functional relations between the rate of decomposition of substrate per unit volume (r_s) or the rate of methane generation (r_n), and the various influential parameters. Without such relations rational and rigorous design is hardly possible.

Attempts to develop these relations have taken two broad routes: to use empirical relations as a basis for correlating the rate with the primary variables; and to base the form of correlation on a more soundly based theoretical model. Although the latter course has much to commend it, it is fraught with difficulties, given the complexity of the process. Consider, for example, the processes involved in the decomposition of cow dung. The feed contains a range of organic materials (carbohydrates, lipids, proteins) with varying degrees of degrada-

Table 31. Kill rates of pathogens during anaerobic fermentation.

Organism — Disease	Temperature (°C)	Retention time (days)	Kill rate (%)
Salmonella spp.	22-37	6-20	82-96
Salmonella typhosa	22-37	6-20	99
Myobacterium tuberculosis	30	n.a.	100
Ascaris lumbricoides	29	15	90
Poliovirus-1	35	2	98.5

bility, solubility, etc. These materials are hydrolyzed, with extracellular enzymes, to a range of simpler organic compounds, which in turn decompose to yet simpler intermediates (volatile acids, hydrogen, etc.). Finally, the methanogenic bacteria are responsible for the last stage of the process — the production of methane and carbon dioxide, which are subsequently released. The problem is thus complicated by: the fact that a large number of chemical species, enzymes, and bacteria are involved in ways which are incompletely understood; the fact that the process involves a complex set of interacting and possibly competing reactions or physical operations; and the possible constraints imposed by limiting reactants, species, or nutrients.

Theories for the fermentation of pure substrates suggest that the rate of substrate utilization should follow the form of the Monod (or Michaelis-Menten) equation

$$r_s = \frac{q_{max}\, S\, x}{K_s + S}$$

where: S = (limiting) substrate concentration; K_s = (half) constant; x = concentration of bacterial cells; q_{max} = maximum substrate utilization rate (per unit cell conc).

For low substrate concentrations

$$r_s \rightarrow \frac{q_{max}\, S\, x}{K_s}$$

and for high concentrations

$$r_s \rightarrow q_{max}\, x$$

When the rate equation is coupled with material balances on the substrate and bacterial matter, overall design equations relating input and output concentrations of substrate to retention time, etc. are obtained. In the case of simple reactions, such modeling procedures are reasonably well established and allow one to interpret results from batch or continuous experiments in a consistent manner (Atkinson 1975). This is, however, not so for the complex situations under consideration here. A further question exists in the case of insoluble substrates (e.g. grass, manure, vegetable wastes): What is the correct 'measure' for S? (the concentration of dry matter, volatile solids?).

There is a good deal of evidence to support the hypothesis that the methanogenic step is the rate-limiting step in the case of soluble substrates; therefore, one can use a Michaelis-Menten form of rate equation to correlate the data. There is no such unanimity when it comes to the more practically interesting substrates such as grass. For example, Chan (1971) fitted data on the continuous fermentation of cellulose to the Michaelis-Menten equation, while concluding that the rate limiting step was the hydrolysis stage. Pfeffer (1968) concluded that the rate limiting step was the methanogenic stage at low detention times (<10 days) and the hydrolysis stage at higher detention times. Hadjitofi (1976) has recently shown that the limiting step is probably the hydrolysis stage at least down to detention times of about 10 days, and that the rate of reaction of the substrate is best correlated by a simple first-order relationship:

$$r_s = -kS$$

while the rate constant, k, followed an Arrhenius relation with temperature

$$k = k_o \exp(-E/RT)$$

In qualitative form this agrees with Boshoff's (1968) results on insoluble substrates using batch reactions, but because so little is known of the dynamics of the process it would be premature to attempt to draw stronger conclusions. It is difficult to gain an unambiguous measure of the bacterial cell (or biomass) concentration with insoluble substrates, but Hadjitofi found that his results did not follow the dependence on x expected from the Michaelis-Menten equation. The reasons for, and implications of, this finding are not yet fully established. Of more direct interest than the rate of substrate removal is the rate of methane production, but there is agreement that the two are related.
Thus:

$$r_m = Y r_s$$

where the coefficient Y depends on the substrate (this can be estimated from first principles with a reasonable degree of accuracy). Some typical results are given in Appendix 2.

A large number of studies report kinetic information in a very simplified form — in terms of the volume of gas produced per kilogram VS added or destroyed (see Appendix 1). (Usually, values range from 0.45 to 0.6 m^3/kg VS added.) A good deal of care should be taken in using these results because many studies give rather incomplete information on the experimental conditions. In addition, there is a good deal of difference between a figure of, say, 0.4 m^3/kg VS added, and 0.4 m^3/kg VS destroyed. The enormous range of operating efficiencies reported earlier should be a warning to all who choose to oversimplify digester performance.

There is a pressing need for further experimental data and interpretation if one is ever to reach the situation of being able to evaluate the optimum set of operating conditions. Because there are differences in the implications of the kinetic models, it is not yet possible to describe accurately the behaviour of a digester over the range of retention times from the minimum (what is it for grass, dung?) upward; nor can one compare quantitatively different digester configurations. Laboratory, pilot, and full-scale experimental trials are required.

Full details of the process are not completely understood and it may well be that the problem is so complex as to defy quantitative analysis (Hobson et al. 1974). The main hope for developing relatively simple models for the rate process is that one of the many stages involved in the reaction set is so slow as to control the overall rate of reaction.

A number of hypotheses have postulated that the rate limiting step is: the initial hydrolysis step (Chan 1971; Hadjitofi 1976); or the methanogenic step (related to the rate of growth of methanogenic bacteria) (Pohland and Ghosh 1971; Andrews 1964; McCarty 1964; Lawrence and McCarty 1969); or the release of carbon dioxide/methane from the bacterial cellular matrix (Finney and Evans 1975).

It might be thought that these considerations are excessively academic but they are not. If, for example, it can be shown that the rate limiting step is the methanogenic step then it could be concluded that the size or physical state of the feed substrate would be unimportant.

Part of the problem in resolving competing claims is that much of the work so far reported has used soluble substrates (glucose, acetic acid), or relatively easily degradable feeds (sewage sludge). Even these studies vary considerably in their conclusions: for example, growth times of methanogenic bacteria are variously estimated at from a few hours to several days. The limiting mechanisms, even with soluble pure substrates, are not unequivocally established. One should not immediately apply results from pure substrates to the fermentation of mixed insoluble substrates because there may well be synergistic or antagonistic effects to alter the picture.

Finally, much of the data contained in the literature is of very dubious value because it is not clear whether the digester ever reached a true acclimatized steady state (requiring 2-3 residence times, or more if a 'seed' is used). It is thus important to record and control all the parameters that can affect digester performance.

Design and Engineering

Earlier discussion described the wide range of operating conditions and efficiencies that are achieved in biogas units. One can conclude that reliable, if conservative, designs exist and it seems very likely that many digester failures are due to either: poor design and construction; or poor operating practice.

Design and construction faults include: poor quality construction; lack of advice/repair backup; blocked inlet/outlet pipes (in bends); sludge buildup in digester; gasholders that cannot be moved/maintained easily; tilting/jamming gasholders; possible washout if flash storms drain through digester; scum accumulation; water accumulation in gas lines; and design close to washout conditions (too low detention time). Bad operating practices include: lack of responsible/knowledgeable people to care for plant; nonexistent operating instructions (especially to deal with faults); insufficient feed material; irregular feed; and lack of maintenance.

There is a need to establish 'good practice' engineering standards in the design, construction, and operation of plants. This could lead to substantial immediate improvements all-around: to date, there has been surprisingly little diffusion of know-how. Monitoring and collecting data over reasonable periods of time would allow accurate comparisons to be made, and would involve measuring at least the items shown in Table 32.

Table 32. For accurate comparisons to be made, at least the following measurements must be completed.

Slurry input and output	Gas output	Other state variables
Dry matter	Gas rate	pH
Total solids	CO_2:CH_4	Temperature
Volatile solids		(mixing — power input)
BOC, COD		
Water content		

A major area for improvement is a reduction in capital cost as a consequence of one or more of: improved efficiency (higher loading, etc.); changes in configuration/materials of construction; or increased thermal efficiency/conservation.

It should also be noted that the technology is susceptible to on-line improvement by regular monitoring of the major variables and by controlled changes in operating conditions (i.e. evolutionary operation, which is a proven method for handling complex systems).

Operation and Control

Perhaps the single most important rule in operating a digester is to attempt to maintain the operating conditions (via the input) as steady as possible. The more frequently and regularly the digester can be fed, the better (Hobson et al. 1975). Air must be excluded completely.

An excellent guide to the operation and maintenance of gobar-gas plants has been produced by Finlay (1976). Apart from setting out the procedures for normal operation, he also gives an invaluable check list for abnormal conditions. (In many conditions the wisest course of action is to leave well enough alone.) It should be standard practice to have documentation/instructions of this type for all operational plants. Apart from mechanical failures, blockages, etc., the main indicators of digester performance are:

Gas production — if this falls steadily, the digester is failing. On the other hand, there are inevitably day-to-day variations in this parameter.

pH and volatile acids — if the volatile acids concentration increases, the process is in danger of becoming unbalanced. Because of the buffering capacity, these changes will not be noted immediately in pH changes; therefore, pH is not a very sensitive indicator.

Alkalinity — bicarbonate alkalinity provides the basic buffering mechanism and if this capacity is reduced to the point where the alkalinity and volatile acids are equivalent, trouble is imminent (alkalinity can be measured by titration).

Smell — normally, the odour of the sludge is not unpleasant, but if conditions are upset, the odour will become unpleasant; however, this is not a very rapid nor sensitive test.

The main methods of controlling/balancing the digester are: (1) *maintaining the bacterial population* — It is not easy to correct for changes in population as exemplified by, for example, a buildup of volatile acids. The pH can be controlled to some extent (Mosey 1974) by liming or other additions; alternatively, a small increase in temperature should promote the methanogenic bacteria at the expense of the acid formers; (2) *uniform feeding* (already commented on); (3) *mixing and time for digestion*; (4) *maintaining uniform and steady temperature*; and (5) *pH control* — If it becomes necessary to add alkali to the digester to raise the pH, it is important not to allow the concentration of any cations to reach toxic levels (McCarty 1964). Chemicals (lime, sodium hydroxide, ammonia) should always be added gradually.

Gas Storage and Handling

The range of end uses for biogas have been discussed. The requirements for various household uses can be estimated on the basis of known consumption rates, and a typical set of data is given in Table 33.

Generally there are few problems of a research and development nature in using biogas. The dangers and limitations in handling are well documented, as is burner design (see Sathianathan 1975). The main comment to make is that great care should be taken not to sacrifice safety or reliability so as to produce cheap burners or stoves. On the other hand, a major source of loss and inefficiency in gas utilization is at the burner, and the design of cheap and efficient burners and stoves should have a high priority.

One problem alluded to earlier is the possibility of water condensing and blocking the gas line. This can be avoided by a simple drain.

Purification

Methods for removing H_2S and CO_2 are well-established (Meynell 1975; Sathianathan 1975) and relatively cheap. Again, the emphasis should be on good practice, and the production of standardized, robust, simple devices.

The main areas for development studies into the technology are: the use in engines (to evaluate performance characteristics over the complete range of interest); cheap distribution systems using local materials; possible end uses of CO_2; and development of cheap, efficient, versatile burners and ovens. All these projects are ones that depend on local needs, priorities, and therefore definition. Similarly, the relative economics of compression, purification, etc. are best handled locally. There are, to repeat, no problems in calculating power requirements and costs.

Table 33. Uses of methane and requirements (from Sathianathan 1975).

Use	Quantity (m^3)	Rate
Cooking	0.32	5-cm diam. burner/h
	0.46	10-cm diam. burner/h
	0.63	15-cm diam. burner/h
	0.28-0.42	per person/day
	0.07	boiling water/litre
Lighting	0.07-0.08	1 mantle lamp/h
	0.14	2 mantle lamps/h
	0.17	3 mantle lamps/h
Refrigerator	1.07	flame operated m^3/h per m^3 refrigerated space
Incubator	0.36-0.71	m^3/h per m^3 incubator space
Gasoline engine[1]		
CH_4	0.42	per kW/h
Biogas	0.60	per kW/h
Equivalent to:		
(a) Gasoline		
CH_4	1.00-1.18	per litre
Biogas	1.33-1.85	per litre
(b) Diesel oil		
CH_4	1.11-1.39	per litre
Biogas	1.48-2.06	per litre

[1]at 25% efficiency.

Instrumentation

A wide range of measurements can be made on an operating digester. The list of measurements given earlier may be taken to be the minimum necessary for evaluation and comparison of digesters.

As far as the operator is concerned, however, the object will clearly be to attempt to achieve stable operation at minimum cost. There are two ways of approaching this problem: (1) Attempt to control, as far as possible, all inputs to the process (i.e. water, waste, temperature of feed, etc.). As emphasized above, the closer is the operation to constant conditions, the better. This can be achieved by proper measurement of the quantities involved, etc.; (2) Monitor the variables most sensitive to digester performance. In practice, the easiest measurement to take is the temperature. As we have seen, pH is not very sensitive, but is clearly a key parameter. A simple indicator test would be extremely useful. Gas composition can be measured easily and fairly accurately (e.g. organic analysis), and the gas production rate, although perhaps not easy to measure directly, can be monitored under constant demand conditions by a skilled operator.

Relatively large-scale operations allow one to monitor more variables, to control the inputs more carefully, and to incorporate more control. As noted earlier, there are very strong arguments in favour of a batch operation because the process is less sensitive than a continuous plant.

Algae and Oxidation Ponds

As discussed earlier, there is a significant potential in the growth and utilization of algae. The use of oxidation ponds (without algal growth) has been discussed, and there is little more to be added here, other than to warn that in tropical climates it is quite possible that algal growth will commence. This may well be a severe embarassment with respect to the major objective of controlling and stabilizing the liquid and solid wastes.

There have been a number of studies of algal growth rates on various substrates (sewage, McGarry 1971; pig manure, Boersma et al. 1975; digester effluents, Obiaș 1976). There is little doubt that processes incorporating algal ponds and subsequent fish ponds are potentially viable. Their economics seem, at present, unclear (see Sumicad 1975). The major danger at this stage is to take 'best condition' figures from a laboratory study and base an assessment of what will happen in a practical situation on this. It is even worse to take even more optimistic figures as an indication of what might happen (see Malynizc 1973). Shaw (1973) gives a good survey of the range of problems and factors affecting algal growth: environmental conditions; pond design; loading; throughput; nutrient supply; and algal population.

Perhaps the single most important problem in the technology is the collection and drying of the algae. Unicellular algae are not easy to filter and collect. There is clearly need here for engineering studies (as a function of algal species) of the methods of algal separation (floatation? flocculation? centrifuging?) and their costs, as well as microbiological/public health studies of the algae and their consequences.

A typical chain of design calculations (excluding financial and economic evaluation) would follow the sequence: specify gas requirement; estimate substrate requirement for range of possible substrates; estimate water requirement; estimate slurry production rate, composition, BOD; choose digester temperature; estimate digester volume, dimension, requirement, materials including adaption to local availability; and estimate energy losses. It is possible to calculate all material and energy flows implicit in this chain of calculations to an accuracy sufficient for preliminary feasibility studies. This is possible for both batch and continuous processes. The material requirements for digester and peripheral equipment manufacture can also be estimated although only approximate design is possible.

Detailed calculation and optimization are not feasible at this stage; however, some of the key areas for further technical work are discussed in the following section.

Research and Development Priorities: Some Suggestions

These comments are tentative suggestions that attempt to summarize in a general way the most useful lines for future work.

(1) The clearest gains to be made in the core technology are in the area of capital reduction — especially by seeking methods of reducing digester and gas-holder volumes and/or by suitable choice of construction materials (see Appendix 3).

(2) It seems that, in general terms, both technical and socioeconomic factors favour larger rather than smaller units (i.e. community level rather than household), and a higher priority would thus be justified with respect to community-scale operations (see following chapter by Andrew Barnett).

(3) Given the wide range of competing possibilities, work should, where possible, be placed firmly in a context that recognizes the existence of alternative solutions or systems, and preferably be related to systems studies of the problem.

(4) A very large proportion of the work needed is actually development and the establishment of good engineering practice rather than research.

(5) It is taken for granted that all laboratory/pilot-scale studies will involve a degree of modeling — that is, relating behaviour to basics and/or dealing with results statistically.

Biogas Technology: A Social and Economic Assessment

Andrew Barnett

In this chapter the problems of assessing the worth of the technical[2] options described previously are discussed. The evaluation of biogas systems presents a number of problems associated with making choices between production techniques that are small in scale and are considered appropriate for use in the rural areas of the Third World. These problems have been discussed (Stewart 1973), but very few of the empirical studies of choices in nonagricultural rural technologies have provided a firm enough base on which major policy decisions might be made (Bhalla 1975; Carr 1976). It is hoped that by addressing the social and economic aspects of biogas technology we will contribute to the more general debate about technical choice at the village level.

The appraisal of biogas technology involves establishing the set of alternatives with which it is to be compared. Biogas plants have to be seen as one of a number of possible uses of (rural) resources, but from the economist's point of view, these other resource uses need not necessarily be connected with energy or fertilizer. If this logic is followed the value of biogas techniques becomes a function of the genuine alternatives that there are to biogas. Much of this chapter is therefore devoted to establishing just what these alternatives might be and what objectives they are to meet.

Once the importance of alternative resources uses is understood, the practicality of biogas plants may be expected to vary between locations; their success will depend on the particular circumstances in which the investment takes place. For example, where an investment in a village is isolated from the rest of the economy by difficult communications or the lack of cash, surpluses and shortages can build up around the project very quickly and these can affect the project either by starving it of inputs or by reducing the value of its output. In a different way, the availability of an alternative source of energy, such as electricity, will vary substantially from location to location and this will also affect the need for (and therefore the value of) a source of energy such as biogas.

It is not only the physical environment that can affect the worth of biogas; the assessment of biogas technology must also be undertaken in the context of the social and economic structure in which it is developed and used. The influence of social structures is therefore the second major

[2]In this paper the terms technique and technology are used somewhat loosely. The term biogas technology is used to signify a collection of hardware, knowledge, and supporting systems associated with the production of methane gas on a scale and at a level of sophistication that is at least theoretically possible in the villages of the Third World. The term includes both those plants that currently exist and those that could exist with the current state of knowledge. The biogas technology referred to here excludes those aspects of the production of methane (other than knowledge) associated with the larger-scale and more sophisticated processes that are currently used in developed countries (usually as part of sewage disposal systems). The term biogas technique is used to describe a particular piece of hardware and is used to distinguish biogas plants of differing scales and designs.

theme of this chapter. Different social groups want different things and they value them accordingly. Social structures regulate how much access individuals have to the capital necessary to use biogas technology, and influence the distribution of the effects that biogas plants produce. The importance of some of these effects is evident when it is realized that small-scale biogas plants have had harmful effects on the distribution of income in certain circumstances.

The third theme of this chapter is the selection of research priorities. This selection is complicated by the sensitivity of biogas to changes in particular village characteristics. In such circumstances, it may be more important to establish an appropriate structure and process for making choices about particular biogas techniques and for determining research priorities, than to establish the research priorities themselves. When the value of any particular change in biogas plants is likely to be so influenced by the location of the plant, research priorities themselves will vary between locations. This means that the social and economic assessment of this particular technology may be more a matter of deciding in conjunction with villagers and engineers which aspects of the technology might be developed to meet a particular set of problems, than evaluating a static set of known techniques for making gas. The technology is currently undergoing considerable change, but so far only a small number of known designs for biogas production have been built and tested. In the future it must be expected that a new set of techniques will evolve that will greatly reduce the costs of biogas; costs will be reduced both by increasing the efficiency of the plants and more importantly by reducing capital costs through the use of new designs and different construction materials.

In these changing circumstances it should be stressed that the current enthusiasm for biogas should not be interpreted as meaning that the technology has already been shown empirically to be the best means of satisfying many of the needs of rural peoples. Nor can the conclusion be justified that biogas has no future without more detailed analysis of the current situation in rural areas and the characteristics of the new designs.

An attempt is made to direct the reader through a range of problems and errors that might be expected in the evaluation of biogas and these are then illustrated by previously published attempts. No attempt is made to provide a 'cookbook' of evaluation procedures.

The analysis of rural technologies can be carried out at various levels of sophistication. There is considerable danger, when attempting to consider a large number of possible problems in a somewhat abstract way, of merely adding to the mystification of the problem and further alienating those who will be affected by the choice of a particular technique. This chapter clearly presupposes a structure in which 'we' try to make decisions about 'them.' This is not the only way. The problem of the choice of a particular technique may well appear much simpler to those that are actually affected by the choice. The problem of development is more one of getting the social structure right rather than one of deciding which particular gadget is to be preferred.

It is assumed that the chapter by Leo Pyle has been read to gain some understanding of the biogas production processes. It is further assumed that the procedures of social cost-benefit analysis are known, or can be learned from the sources quoted in this chapter.

The chapter is divided into five sections: a general approach setting out the framework in which biogas technology can be analyzed; a valuation of common inputs and outputs of biogas plants; five case studies of attempts to carry out social and economic evaluations of biogas plants; the social and economic determinants of the demand for biogas; and an approach to research priorities.

The General Approach

The primary need is for a logical and consistent framework in which the problem of the evaluation of investments in biogas can be analyzed and appropriate questions can be examined. Such a framework should serve two main purposes: it should make explicit the assumptions that have to be made in the analysis and it should force the evaluator to examine the full range of possible alternatives.

The evaluation of biogas investments can either be approached as a macroproblem, setting the investments in the wider context of the economy's overall fuel and rural development policies, or it can be treated as a microproblem, in which the returns to a single investment are examined at a specific location and within a specific set of macroconditions. Clearly macro and microlevels interact, in that the macrodecisions depend to an extent on information at the microlevel on the viability of the individual investments. A review of the literature on the evaluation of biogas systems shows that these microdata are not yet available; this is either because of the imprecise nature of the data used in the few analyses that do exist or because of the difficulties experienced in successfully running the existing plants. In addition to this, the viability of an investment at the village level (such as biogas) depends crucially on the particular characteristics of the location of each investment. It is therefore important to first establish which village characteristics have the most influence on the viability of the biogas investment and from this generalize to the more macrolevel about the distribution of these characteristics throughout the country. For these reasons it is suggested that the problem of the evaluation of biogas should be treated, initially at least, as a microproblem.

The most widely used logical framework for the evaluation of microinvestment decisions is social cost-benefit analysis (SCBA). Considerable advances have been made in recent years in refining the logical consistency of these appraisal systems and it is recommended that at least one of these systems is used for the evaluation of biogas. The most recent manual has been produced by the World Bank (Squire and van der Tak 1975). This is a particularly clear, if somewhat condensed, version of the approach (see also UNIDO 1972; Little and Mirrlees 1974; Irvin 1976).

The approach adopted by all the recent manuals of SCBA is predominantly economic, but this is only one of a number of possible dimensions against which the impact of an investment can be judged. What is suggested, therefore, is that a 'softer' form of SCBA be adopted, using the framework set out in the various manuals as a guide, but taking into account a greater range of possible social and environmental effects and attempting to give sufficient weight to those effects that cannot be precisely measured on a scale such as that provided by money values. This 'softness' does not necessarily imply a weakening of the overall logic of the analysis but reflects the reality that many important events are neither economic nor can they be precisely quantified.

The cost-benefit approach attempts to determine the physical relationship between inputs and outputs associated with a particular investment and then places economic and social values on these events. It is in essence a process for weighing the various characteristics of alternative courses of action — and as such is the decision process of everyday life.

The Political Framework

Investment decisions and the assessment of the costs and benefits that result from the investment are primarily political decisions influenced by two factors: the nature of the group making the decision, and the social and economic structure of the society. It is a political process because the decision-makers are forced to make an explicit choice about an investment that helps one group of people rather than another. The position of the decision-makers in the social structure will strongly influence their views on how investment should be used to further development objectives. These views may differ drastically from the views of the people that will be affected by the investment, and more importantly may fail to take into consideration the way in which the existing structure will affect the actual distribution of costs and benefits.

The social structure influences the distribution of costs and benefits among social groups in a number of ways: for instance, where some of the factors of production are monopolized by a particular group, the introduction of a new technology, however beneficial to the individual owner, may merely raise the amount of surplus that can be expropriated by the monopolists. Conversely, investments in new technologies can also be used to alter the existing distribution of power by helping to break dependent relations.

All too often the choice of techniques and particularly the advocation of 'appropriate' technologies is abstracted from the realities of political and social structure (Cooper 1973). The introduction of small-scale biogas plants, for instance, may have the effect of fostering individual actions rather than cooperative action; it may satisfy the needs of only those who can afford plants, thus reducing the pressure for more redistributive solutions; it may increase the value of inputs such as cow dung and subsequently deprive the poorer sections of society of its use. Similar problems were well documented following the introduction of high-yielding varieties of food grains during the so-called Green Revolution in India (Griffin 1972).

It is hoped that the approach to the evaluation of biogas plants suggested here will help to make explicit the choices that have to be made and the political nature of these choices.

The Examination of Alternatives

The evaluation of the impact of an investment is, in principle, the comparison of the situation 'with the investment' and the situation 'with the next best alternative investment.' This concept of 'opportunity cost' is crucial to the approach of social cost-benefit analysis and is described in detail later.

At this stage the important point is to decide what realistic alternatives there are to the investment in biogas: what is to be compared with the biogas system?

In a microanalysis such as the one proposed here, the 'next best alternative investment' is likely to be another investment in the village: such an investment might be in an irrigation pump, in land drainage, in paying off previous debts, in buying new land etc. From the standpoint of the whole economy the 'next best alternative investment' might encompass a wider range of activities including the production of fuels and fertilizer by other (possibly larger scale) processes. The set of possible alternatives can obviously be very large and depends both on who controls the available investment resources and on the characteristics of the economy.

Some evaluations of biogas take a much narrower view than the one implied by SCBA and consider only a number of different techniques for producing methane by anaerobic digestion. It is quite clear that there are many possible designs and scales for biogas production that have not yet been built or tested. The appraisal of these designs is an important task, but it is equally important to consider how much the products (methane gas and slurry) are required and whether this is the best possible use of the resources involved.

An improvement on this narrow approach is to evaluate the techniques for methane production in relation to the existing means of satisfying the needs for fuel and fertilizer (this is the approach adopted by the Indian Council for Agricultural Research, Government of India 1976). But this does not answer the more fundamental question of whether investment in methane is the best use of the available resources.

To expand the range of options that might be compared with biogas a useful approach is to consider the range of functions that the biogas plant might achieve. Biogas is advocated on a number of grounds each of which can be achieved by other more or less 'good' techniques. Some of the options are discussed below.

(1) Biogas can be seen as providing a fuel and can therefore be evaluated in terms of its ability to meet some of the villages' energy needs in comparison with other sources of energy. The comparison might include: (a)

firewood, which in many areas is becoming increasingly time-consuming to collect because of its scarcity[3] (Makhijani 1976); (b) electricity, which is not usually used for cooking but can have very low marginal costs where there is surplus capacity in existing generating capacity and the village is close to existing power transmission lines (Prasad et al. 1974); and (c) at a different level, the comparison can be legitimately made with alternative practices associated with the use of existing fuels; for instance, considerably less wood might be consumed if the design of stoves were made more efficient (Makhijani 1976, p. 27).

(2) Biogas plants provide fertilizer in the form of the spent slurry and might therefore be compared with aerobic composting processes or the provision of chemical fertilizer. (Disney 1976 compares biogas with urea production in India.)

(3) Biogas has been advocated as a substitute for other activities that are considered harmful or wasteful, such as the burning of dung and wood. The dung might be better used as fertilizer, and the burning of wood has resulted in deforestation of some areas, which in turn, has led to erosion and flooding (Republic of Korea 1975).[4] The comparison here might be with composting or the growing of trees and other plants (such as water hyacinth) for fuel (Makhijani 1975, p. 114-124).

(4) Another function for the biogas plant is the safe disposal of human (Sathianathan 1975, p. 158) and animal (usually pig) manure (Solley and Yarrow 1975); here the alternatives for comparison might again be composting processes or more conventional waste disposal through lagoons and septic tanks.

(5) Investment in biogas might alternatively be seen as a means of utilizing village resources that are currently going to 'waste' (or at least are being underutilized). This way of looking at the problems gives certain insights and is another formulation of the more general economic problem of the optimum use of all resources.

It is essential that the researcher specifies precisely which options are to be considered and justifies the exclusion of others. It would certainly be unfortunate if the impetus that now surrounds biogas were used to divert attention from the modification and utilization of the mass of other applications of technology at the village level that would appear to produce at least as promising returns as biogas.

However, the fact that biogas is currently fashionable and is being promoted in a number of countries is sufficient justification to modify what would appear, at the moment at least, to be a rather expensive and unreliable set of techniques. But this concern must not overshadow the more general search for the 'best' use of village resources; it does not matter how good the methods of social and economic evaluation are if they are applied to the wrong set of alternatives.

Once the logical framework has been adopted and the broad set of alternatives established, the next stage can be broken down into two tasks: the identification and estimation of the physical quantities involved as inputs to and outputs from the project; and the placing of social and economic values on these quantities. This division is useful in terms of exposition because it helps to make explicit the various assumptions that have to be made, and because many of the errors are made in the estimation of the inputs and outputs rather than in the process of evaluation, which is commonly considered to be most difficult part of the analysis.

---

[3]Makhijani 1976, p. 26; wood, straw etc. have been a traditional source of fuel in rural areas in many developing countries. Makhijani suggests that there is considerable evidence that collecting wood for fuel now requires an increasing amount of labour time. In rural India, anywhere from 50 to 200 or more days of work per family are now required to ensure adequate fuel supplies.

[4]Republic of Korea 1975, p. 2. Forest products and straw account for 92% of Korea's rural fuel needs. Soil erosion, resulting from the denudation of hillsides and the reduction of soil fertility associated with burning straw, are a major problem in Korea.

Physical Input and Output Relations

As an initial stage the inputs and outputs directly connected with the investment have to be listed and the relationships among them established. The establishment of these relationships is the work of engineers,[5] but it is necessary to stress here that a large source of uncertainty and error in the analysis of biogas techniques arises in the estimation of these relationships under normal working conditions. The evidence in the literature is often unclear and a large range of values can be found for most of the essential input/output relations (see chapter by Pyle). Biogas plants, as currently designed and used do not appear to be very reliable (see Table 20), and the actual quantity of gas produced would appear to be considerably less than the 'design capacity' sometimes used in the cost-benefit analyses.

Three Dimensions

The impact of the biogas investment and its alternatives can be compared against a number of possible 'dimensions.' Usually, the only dimension that is considered involves the direct technical inputs and outputs of the biogas plant and the subsequent economic analysis of these quantities. It is advocated that two other dimensions be considered to give a broader view of the impacts of the investment. In addition to the technical/economic dimension, consideration might be given to impacts along the social and environmental dimensions. The content of these two additional dimensions must be specified in some detail and relevant 'indicators' of impacts defined. The problems of specification are quite distinct from the problems associated with the social and economic valuation of impacts once they have been established in physical terms. Values are involved to the extent that different social groups consider certain impacts more worthy of estimation than others, but the problem of specification mentioned here is a question of the kinds of impacts that are to be considered (and measured in such physical terms as kilograms, number of people affected etc.)

The kinds of impacts that might be included in the social and environmental dimensions were discussed at a recent conference of the United Nations Environment Programme (1976). The social and environmental dimensions were each broken down into two strands. The social dimension was related to impacts associated either with structural development or cultural compatibility; the environmental impacts were related to either the quality of human life or ecological balances. Structural development discriminates among the impacts of alternative investments on the basis of the degree to which they promote self-reliance, involve public participation in decision-making and implementation, or reduce dependence at individual, village, and national levels. Other impacts associated with structural development are the reduction in inequalities among groups and individuals in terms of the distribution of needs such as consumption, education, work, and power.

Cultural compatibility examines investment alternatives in terms of whether they build on the endogenous traditions of the society or whether they run counter to them.

The quality of human life aspects of the environmental dimension incorporate those characteristics of projects that satisfy such needs as the need for creativity and the need for local initiatives. Additionally, it lays emphasis on the replacing of one type of activity (such as boring, hard, dirty, repetitive, degrading work) with another. The impacts related to ecological balances include the extent to which nonrenewable natural resources are used, the extent of pollution, etc.

A number of these impacts have been incorporated in previous analyses of the technical/economic dimension. Impacts that have been included in this way are employment effects, effects on the

[5]Moulik and Srivastava (1975) suggested that "about 71% of plant owners experienced technical problems. A large number of plants in the sample were closed due to these problems" (p. 120).

distribution of income, and the costs of pollution. The argument here is that there is considerable value in not trying to load too much on the economic analysis, particularly at the second stage where the economic and social values are ascribed. The incorporation of many separate impacts into the economic dimension tends to shift the balance of power over decision-making in favour of the project appraiser and away from those who will be affected by the investment. Clear problems of employment become reduced to esoteric discussions of shadow wage rates (Squire and van der Tak 1975, p. 29). The separation of impacts into three dimensions reduces the number of assumptions that have to be made by the project analyst and provides a useful checklist of impacts that must be examined for each investment alternative.

A large number of other impacts might just as well have been included instead of the brief list suggested in the UNEP paper. A number of alternative suggestions, particularly along the environmental dimension can be found in the work of the International Council of Scientific Unions (1975) (SCOPE), and Marstrand et al. (1974) (SPRU). An attempt to incorporate more of the social dimension is to be found in Szekely (1975).

The point is that a wider range of physical impacts than is usually implied by economic analysis should be considered in the evaluation of new production techniques. The problem of integrating these impacts into the decision-making process is not necessarily simple and will be discussed later.

Quantification

The selection of impact indicators is clearly a crucial stage in the appraisal process, and it is apparent that of the many possible indicators there is considerable variation in the extent to which they can be measured. The highest form of measurement (cardinal scales) in which it is possible to say that an impact is so many times greater or smaller than another, can be ascribed to many of the impacts and these are often the core of the economic/technical analysis. But other impacts have to be measured on less powerful scales — where the impact can only be said to be greater or less than another or, at an even lower level, where the impact can merely be seen to exist or not to exist. Many choices in everyday life are made on the basis of data that cannot be precisely measured, but there is a tendency in much (economic) project analysis to pay more attention to precisely quantified data to the exclusion of all else. This bias in quantification sets up a corresponding bias when social and economic values are placed on these impacts — higher value is placed on the more measurable aspects of the problem. The procedure suggested in the section entitled "valuation" allows for the inclusion of data that can only be quantified on less powerful scales.

Limits

The emphasis so far has been on trying to expand the range of problems that might be examined and the factors that might be taken into account. At some point, however, limits have to be placed on the problem to define its boundaries.

Any investment takes place as part of a subsystem of events, and these subsystems connect with other subsystems. In the case of village technologies, these systems are often compact and the alteration of one part of the system considerably affects another. This compactness poses particular problems in the appraisal of rural projects if the project is partially isolated from a 'market' that can easily dispose of surplus outputs that are created and supply needed scarce inputs. This isolation stems from the costs of transport and travel and from the relatively large proportion of rural activities that are not monetized. In such situations the usual assumptions of microproject appraisal that the project is 'marginal' to the rest of the economy cannot be made, in the sense that its existence will not have an effect on prices (particularly 'world prices') (Squire and van der Tak 1975, p. 32). While such interrelatedness is a problem for SCBA, it may be of considerable benefit to rural peoples as the by-products of one village activity become the very cheap inputs of another.

The usefulness of the various appraisals of biogas largely hinges on the extent to which these important effects within the system have been included. Apart from the impact on the social and environmental dimensions, which are often not considered, the analyses often disregard those effects that take place in part of the system not immediately surrounding the project. Such effects have been called 'second round' or 'linkage' effects. An example of these types of effects occurs when the output of the project under consideration increases the supply of a particular commodity sufficiently to reduce its price within the locality. Other activities that in turn use this commodity as an input will be affected beneficially by this reduction in price. These benefits, which occur in the 'second' round of transactions within the system, or which induce additional 'linked' investment further down the chain of the system, can legitimately be included as benefits to the project being appraised. Similar effects can also arise in connection with the project's need for inputs. An increase in the demand for cow dung, which might result from the introduction of a biogas plant, can have the second round effect of reducing the availability of cow dung to other existing users further 'up' the chain. Improvements in the efficiency with which wood is burned may not only have beneficial ecological effects in terms of deforestation and erosion, but it can also have harmful second round effects on those people whose sole source of income is the collection and sale of firewood (such a situation might arise with the wood-collecting 'tribals' of western Maharashtra in India).

The question arises as to how many of these effects it is necessary (or cost-effective) to examine. In the evaluation of investments within a single commercial venture, it is often the case that (good or bad) effects that do not affect the firm itself, because they do not take place within the physical confines of the enterprise, need not be considered. These effects are described as 'externalities' because they are external to the enterprise. In such analyses the limit to the number of effects that must be examined is clearly defined and is largely reflected by the items in the firm's accounts. In social cost-benefit analysis no such clear limit (either of geography or system) exists and there cannot be a clear set of 'external' effects.

By their nature there cannot be a firm set of rules to determine which effects should be considered in social cost-benefit analysis. The choice of a cut-off point, beyond which effects need not be considered, is a matter of judgement and experience; judgement about the likely size of possible effects — the smaller the effect the less its exclusion matters from the microanalysis — and experience of the kinds of effects that have been encountered with previous investments. The choice and specification of the different dimensions along which projects are to be evaluated have a certain role in this respect because they draw the attention of the project analyst to a range of possible effects. The application of social cost-benefit analysis in villages would seem to be less a problem of the rigorous application of a set of economic techniques, but more one of understanding the variability of the context in which the investment is taking place.

Three points emerge from this discussion about the appraisal of biogas investments.

First, it is vitally important to examine the appropriate system surrounding the project. This includes examining the chain along which inputs will actually arrive at the project and along which the outputs will proceed after it. It also involves ensuring that similar 'levels' of system are being compared in the examination of alternatives. A common error is of the type in which the cost of fertilizer or fuel from biogas is compared with the cost of similar outputs of larger-scale production techniques. However, no account is taken of the costs associated with the delivery of the output of the large-scale unit to the point of consumption (the village);[6] these are the transmission costs of electricity or the delivery cost of fertilizer or kerosene. It may be useful to use flow charts to represent the systems being compared.

Second, as the choice is always between alternatives, it may be that the second round effects will be of a similar form and size whichever investment is made in a specific location. To the extent that this is true these effects can be excluded from the analysis without affecting the decision as to the best investment. However, in practice this logical nicety is more usually the refuge of the self-justifying analyst!

Third, the importance of these second round effects will depend on how large the investment is. Even if the investment is quite small, the introduction of many such investments (as with the widespread introduction of biogas plants) may well add up to the sort of nonmarginal change that can alter many of the parameters of even a highly responsive 'market' system and make microproject appraisal difficult to apply. This raises the question of how reliable a few micro cost-benefit studies of biogas are likely to be for deciding on such nonmarginal changes as the introduction of many thousands of biogas plants in high concentrations. The 'whole' is obviously going to be greater than the sum of the 'parts.'

Again, a compromise must be reached between the ease of analysis and the extent to which the analysis describes the complex reality. What can be said is that the compromise should not favour economic rather that the social and environmental dimensions, nor should it favour the quantifiable in preference to the unquantifiable.

Reliability of the System

The estimation of the physical input/output relations associated with biogas investment must also take into account the reliability of the system and its robustness. A number of factors which influence the production of gas are discussed in the chapter by Pyle, and include air temperature, shock loading, poisons, and variations in inputs. Where possible, 'expected' values should be used in which each possible level of output is weighted by the probability of its occurrence as described by Squire and van der Tak (1975, p. 44). More generally, it is important to consider whether changes in design or operating procedure, which increase the amount of gas produced, alter the reliability of the system. It is likely that a tradeoff exists between increases in gas output volumes and the reliability of the system.

The lack of reliability of the system in providing adequate supplies of fuel has been quoted as a major reason for the non-acceptance of methane generators.[7] Where this problem is important, the physical resources required to provide a backup source of fuel (and cooking stoves etc.) should be included as additional costs for the biogas system.

Valuation

The various inputs and outputs identified on the technical, social, and environmental dimensions have to be valued. Valuation is made explicitly and implicitly in terms of a set of objectives, and these objectives are likely to vary from person to person and among different social groups. The sorts of groups whose objectives might be considered in the evaluation of village technologies are: (1) the government (as a proxy for society?); (2) the owners of the investment — the farmers; (3) the village as a whole; and (4) the various elements within the village: large farmers, other landowners, landless people, women, etc. The actual choice will depend on the local circumstances and the distribution of the effects of the investment among these groups.

---

[6]This is particularly true when the only cost allocated for a competing product is its unit production cost. See for instance, Government of India (ICAR) 1976, p. 28, where no cost of transporting kerosene or fertilizer is included.

[7]Singh (1976) contends that operational difficulties are the single most important negative factor militating against the acceptance of biogas by individual families. See also Moulik and Srivastava (1975).

Each of these groups might be expected to consider different things important and to value them differently. The most widespread system of valuation might be termed financial analysis, but it is clearly only one of many possible systems. Financial analysis is usually carried out using the existing set of market prices to value impacts, and the objective chosen is usually related to the maximization of profits over some specified time or the maintenance of some minimum acceptable income level. Such analysis is normally conducted in terms of the effects on the investor alone.

Social cost-benefit analysis provides an alternative system of values and has usually been carried out from the point of view of the costs and benefits to government. The objectives have been largely economic, involving some combination of such considerations as the rate of growth of the economy (and therefore the amount of investable surplus arising from the investment and the level of employment) and the distribution of consumption among various groups. Such analysis takes as a starting point that market prices do not necessarily reflect the values of society.

The essential problem of valuation is to find some way of combining the values placed on the various impacts so as to reduce them to some manageable aggregate. In this way, the 'costs' can be set against the 'benefits' by the decision-makers. The use of money values as a basis on which to make these valuations is attractive: this is the basis on which many choices in everyday life are made and it is widely understood. It is normal to take market prices as a starting point for the aggregation of impacts into costs and benefits, but two fundamental reservations must be made.

Market prices do not necessarily reflect the valuations that a particular group might place on project impacts. The reasons for this are discussed at great length in the literature (Squire and van der Tak 1975, p. 15-18), but it is clear that market prices reflect the current distribution of income and production, which may well be considered to be unsatisfactory. Prices are also the result of so-called 'market imperfections' such as monopoly elements, lack of knowledge by buyers and sellers, transportation costs etc. For these reasons adjustments have to be made in the analysis to market-price valuations. These adjusted prices are often termed 'shadow prices' or 'accounting prices' and have been defined as "the value of the contribution to the country's basic socio-economic objectives made by any marginal change in the availability of commodities or factors of production" (Squire and van der Tak 1975, p. 26). Such a definition need not necessarily relate to the 'country's objectives' but can relate to any specified group within the country. The precise procedures for constructing shadow prices have now become well established, and they are described in considerable detail in the project appraisal manual mentioned in the first section of this paper.

It should be stressed, however, that all the systems described in SCBA manuals utilize a government objective related to the rate of growth of the economy. This objective is sometimes modified by consideration of the need to trade off economic growth for greater employment or for a more equal distribution of consumption. Costs and benefits are therefore calculated in terms of their contribution to, or savings of, consumption (as in the UNIDO Guidelines) or investment (as in Little and Mirrlees, and Squire and van der Tak). A number of ways of modifying shadow prices to reflect other subobjectives are discussed in the literature; for instance, the placing of higher value on consumption that goes to satisfy the basic human needs of underprivileged groups. Certain social and environmental effects can also be easily incorporated in this analysis as costs or benefits: pollution damage as a cost, and increased consumption of luxury items valued at zero, etc. But, as argued above, there is considerable value in keeping these subobjectives separate.

A number of impacts are not usefully valued in 'money' terms, and certain objectives are more easily understood when not aggregated with economic objectives. The impacts of alternative investments can be displayed in relation to a number of separate criteria with no attempt made to aggregate the impacts into a single index of net benefit. It should be possible to construct a matrix showing the impacts in terms of various criteria for all the investments considered (see Fig. 23).

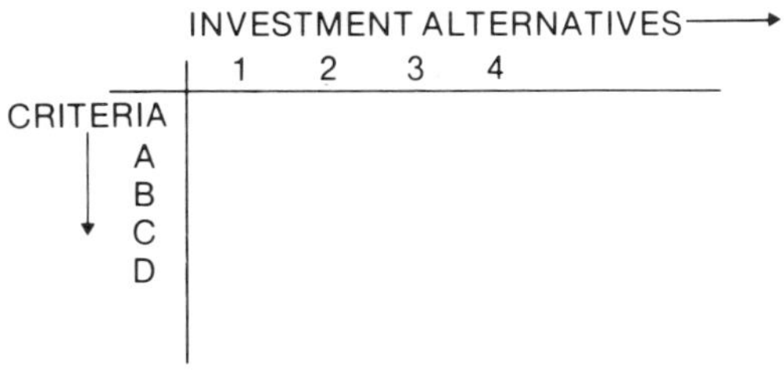

Fig. 23 A simple matrix of the various investment alternatives and the criteria governing their choice can be a very useful tool.

Each cell of the matrix contains a gradation good/neutral/bad (+, 0, -) or the appropriate score on a cardinal scale, such as rate of return, net present value, etc. If need be, some of the criteria can be given infinite weights such that if the project has a negative impact on a particular criterion, this outweighs any possible advantages on all the other criteria. Such an infinite weighting might be given to certain pollution standards or to the worsening of income distribution.

The problem of adopting such a matrix displaying the possible effects of investment choices is that difficulties arise in trying to trade off one criterion against another. For instance, how is the choice to be made between one investment that scores high on one criterion and low on another and an alternative investment that has exactly the opposite characteristics? First, it is rarely the case that decision-makers can specify in advance the precise (mathematical) weights that they would give to different criteria (such weights can be estimated in retrospect on the basis of previous decisions, but there is no reason why the decision-makers should consistently stick to such a valuation, UNIDO 1972, Chapter 18). Second, the data can be presented in such a way that they show how much of the score on one criterion has to be given up to gain increases in another; for instance, the reduction in the investable surplus generated by the project that would have to be given up to increase the level of employment.

Again no hard and fast rule can be adopted. It seems that in many cases the data are sufficient, when presented in the matrix form, for decisions to be made — or at least for areas to be identified in which greater elaboration is required. This is particularly true when the number of options and the number of criteria are relatively small. A matrix of ten investment alternatives and up to five criteria would be a considerable improvement on the level of information on which many decisions are currently based; yet, it would be sufficiently small for the comparisons to be made by eye. A certain degree of aggregation cannot be avoided and a choice has to be made between the need to present the largest amount of information that can be usefully handled, and the need not to bias the result by attaching inappropriate weights to the various impacts. A number of environmental impacts for instance might usefully be aggregated into a single scale ranging from +5 to -5.

Possible criteria that might be used in a decision matrix might include: the financial (money) returns of the project to particular groups (the owner, poor sections etc); the net present value of the social returns (from the SCBA) to the government and other groups; an indicator of the relative effects of the projects on the distribution of income; an indicator of the employment generated per unit of capital invested (or per unit of some other scarce resource); an indicator of the damage to the environment, if any; and an indicator of the contribution to the village's self-reliance. The selection of the criteria will depend on who is making the decision: for a particular group many of these criteria will be unnecessary and a decision will be more clear cut.

Such criteria do not have to be used in the purely passive function of screening out investment alternatives; they can also be used actively to define the kind of technology that is required. With the bias in the current distribution of research and development expenditure in the world, it is likely that many current village level technologies are less developed than the technologies used in urban and richer societies. Therefore, the evaluation of the existing range of techniques may well show the more developed 'western' techniques to be superior. But, it would seem possible to design techniques for use at the village level that would satisfy a number of social objectives and also produce high social and financial returns.

Valuation of Common Inputs and Outputs

The general idea behind the valuation suggested here is that of 'opportunity cost.' That is, all inputs and outputs of the project being appraised should be valued in terms of the loss to the chosen objective that would have resulted had they been put to their next best alternative use rather than in the project being appraised. This principle is of crucial importance, particularly where many of the items in the evaluation are not traded.

The procedure involves deciding what real alternatives there are to the use of particular inputs and outputs. In villages, the real alternatives (rather than fanciful ones) are often difficult to establish, but the 'alternative use' actually chosen considerably affects the viability of each investment. It is for this reason that the conclusions about the feasibility of a scheme in one location are often difficult to generalize to another.

The principles of opportunity cost valuation will become more apparent in the discussion of the following five broad categories of inputs and outputs associated with biogas: cellulosic organic material such as dung inputs and slurry outputs; methane gas; labour; capital; and other outputs such as the improvements in public health.

Cellulosic Organic Material

There would seem to be five possible opportunity costs for the valuation of these inputs.

Valuation According to Market Prices

In certain areas cow dung is bought and sold, and this price might be an indicator of the return that could be achieved from other uses of dung. The argument is that dung would be sold for more money if the market price underestimated the uses to which it could be put, and would have no sales value at all if the market price exceeded the return that could be expected. However, it is known that cow dung is only partially exchanged for money — in India only 2% is traded according to an ICAR Report (Government of India 1976, p. 2), and less than 5% is sold in Pakistan (Government of Pakistan 1969, p. 67). Therefore, the dung that is traded may not be representative of all the dung available. This may be because those who sell dung may place a low value on it because they have no land on which to use it as fertilizer and cannot use it all as a fuel; conversely those wishing to buy the dung might have no money.

Valuation in Terms of the Use of Dung as a Fertilizer

In this valuation it is argued that the next best thing to putting the dung into the methane generator is to put it directly on the fields. The opportunity cost of the dung when used in the generator is the loss of the fertilizer value of the dung when used on the fields. Two points need to be emphasized. First, the dung will also have a fertilizer value once it has been passed through the generator (i.e. as slurry), and although there is no reason to believe that the fertilizer value of dung will equal that for slurry (some studies show a superiority for the slurry, Sathianathan 1975, p. 81-83; also see case study 1) the fact that similar procedures for valuation are used means that any errors that are made on the cost side in the valuation of the dung will be off-set to some extent by similar errors on the benefit side in the valuation of the slurry. Second, the fertilizer value of the dung might be changed if it were aerobically composted before being put on the land. This emphasizes that it is important to consider the net cost of the dung to the methane project. If the dung is put directly into the generator, certain savings might be made in terms of the labour and other costs associated with the composting process. The fertilizer value of the compost is counted as a cost to the project and the savings resulting from not having to do the composting are counted as benefits.

The value of the dung (and the spent slurry) as fertilizer can be obtained in two ways. The more usual method is to establish the content in the dung of the chemicals that are useful to plant growth (N, P, K) and value these at the farm-gate cost (i.e. including transport etc.) of an equivalent amount of factory produced chemical fertilizer. A number of problems arise with this method of valuation. First, it cannot be assumed that the plant nutrients contained in the dung can be quantified with certainty, as considerable variation is indicated in the literature. Second, the price of the factory produced chemicals may bear no relation to the 'social' costs of production. This may be because of taxes and subsidies (which can be easily taken into account) or because the market cost of production does not reflect the 'social' costs of production. The points made earlier about the inadequacy of market prices apply equally here, but the point that has particular relevance is that the foreign exchange cost of the product might not be sufficiently reflected in the price. The manual of project appraisal that has been recommended (Squire and van der Tak 1975) approaches this problem by valuing such products in terms of 'world' (or more precisely import/export) prices. Other approaches attach a special shadow price weighting to the foreign exchange component of the costs. A third problem with this approach is the assumption that the beneficial effects of dung, compost, or slurry are equivalent to their content of (chemical) plant nutrients such as nitrogen, potassium, and phosphorus. This is clearly not the case. These 'organic' fertilizers also have the added effect on soil structure of adding humus, which increases moisture retention in the soil and helps prevent soil erosion etc. If these differences between organic and chemical fertilizer are significant, either an adjustment can be made in the value of the dung or these benefits can be picked up by the second method of obtaining the fertilizer value of dung.

The fertilizer value of dung and slurry can alternatively be estimated in terms of the net increase in crop output resulting from their use. This is, in principle, the most correct formulation of the opportunity cost of the dung (or slurry), but it may also be the most difficult to establish. Not only are there the problems of establishing the physical increase in crop output due solely to the dung, but theoretically it is also necessary to establish the correct shadow price for the crop.

Valuation in Terms of Use as a Fuel

The amount of dung used as fuel varies considerably from region to region and among the estimates of the various researchers. The National Council for Applied Economic Research put the figure for India at 22%; the highest figure is given for Bihar State with 60% of the dung diverted for fuel purposes.[8] The heat value of the dung can be estimated, together with the amount of usable heat that will be obtained in the current stoves, and this can be related to the cost of providing an alternative fuel (kerosene, electricity if this were used for cooking, wood, or coal).

Valuation in Terms of a 'Free Good'

This is a situation in which the input was previously unused. This would be most unlikely with dung, but in certain situations it might exist with crop wastes. In this situation the opportunity cost might be thought to be close to zero (with the only

[8]Government of India (ICAR) (1976, p. 10), Government of Pakistan (1969, p. 65), suggest that less than 4% of collected manure is used as a fuel. Berger (1976, p. 11) suggests that in both Nepal and Korea, wood is the major source of fuel in rural areas. See also Mardon (1976, p. 16).

cost being the costs of collection). But this raises a fundamental issue within this sort of analysis: although the crop wastes were not being used, they probably could have been. Should the opportunity cost therefore be zero or the net benefit foregone had the 'waste' been used more productively? This is perhaps a complicated way of saying that if a commodity had previously been unused it is worth considering the full range of possible uses for it before throwing it into the methane generator!

Valuation in Terms of a 'Nuisance'

This is a situation in which the dung or other waste was previously a nuisance and had to be removed or treated at a cost. This can arise with intensive animal rearing where the dung is seen as an effluent problem rather than as an asset[9]. In such a case, the 'cost' of using the dung in the methane generator would, in fact, enter into the analysis as a benefit (i.e. as a negative cost — a cost with the opposite sign to all the other costs).

Methane Gas

The same principles for valuation apply here as with the dung; indeed it should now be clear that the difference between a cost and a benefit, between an input and an output, is merely the sign (plus or minus). The gas is valued in relation to the real cost of alternative energy sources (including the cost of supply). It is worth noting, that if the alternative energy source were electricity (electricity is rarely used for cooking but clearly is an alternative if the biogas is to be used for lighting or motive power), there is a question as to what price should be used for electricity. Apart from subsidies,[10] a characteristic of electricity generation is huge economies of scale combined with considerable 'lumpiness' in the size of possible investments. This means that the marginal cost of supplying electricity from a plant that already exists, but is operating at less than full capacity, might be extremely small — merely the marginal costs of the power plant and the additional transmission equipment. Where villages are close to existing main-line grids, the cost of electricity might be very low indeed (though the price charged by government is likely to equate to the average cost — illustrating a difference between the private and social returns to such a plan). A similar problem arises when the price of electricity is calculated on the basis of a new plant, the oldest plant, or some sort of average of the two. It is to be expected that these prices will vary considerably.

If the alternative to biogas is wood (and ICAR suggests that for India wood satisfies 58.6% of rural fuel needs, Government of India (ICAR) 1976, p. 9; Berger 1976, p. 11; Mardon 1976, p. 16) the cost in terms of the opportunity cost of the labour required to collect the wood might considerably exceed any other valuation such as the market price or the thermal equivalent of another fuel. It has even been suggested that the time taken to collect wood in some areas has changed in recent years from a minor chore to the full-time occupation of individual members of the family (Makhijani 1976, p. 26). The other cost associated with the use of wood is that of deforestation leading to erosion and flooding (Republic of Korea 1975, p. 2). Theoretically, it might be possible to establish the loss of crops resulting from the erosion, or the cost of making good the damage done by erosion. But, in most analyses of this sort some arbitrary additional weighting is applied to the cost of wood to see what effect it has on the viability of the various schemes (conversely it might be useful to show what weighting would have to be put on the cost of wood in order to make the biogas plant compete with other processes for the production of fertilizer and fuel).

---

[9]Such is the case in Fiji and other Pacific countries where the sanitary disposal of wastes from intensive commercial piggeries is necessary (see Solly and Yarrow 1975).

[10]Makhijani and Poole (1975, p. 98). The actual cost of electricity to a large Indian village is about 8-10 US cents per KWH. However, the rural customers are only charged about 2 cents because of government subsidies.

Certain costs associated with the use of fuel before the introduction of the methane investment, such as the cooking stove, are 'sunk costs' that would have been undertaken whether the investment in biogas is made or not. (For a discussion of sunk costs see Squire and van der Tak 1975, p. 21.) Such costs should theoretically not appear as part of the costs associated with the alternatives to biogas; the comparison is between the future fixed and variable costs of biogas production and the future fixed and variable costs of the alternative (but also see previous section "The Reliability of the System").

In certain circumstances the value of the gas might be assumed to be equal to the cost of production (less the value of fertilizer). This valuation might be used when the comparison is with some other form of energy.

If the CO_2 produced in conjunction with the methane is separated in a usable form, this should also be valued in terms of its net benefit further down the chain of investments. If the CO_2 is used to increase plant growth in greenhouses, its value is the value of the increased crops minus the cost of separating the gas and delivering it to the greenhouse.

The opportunities for using the gas might well be increased if the gas can be compressed, stored, and transported. The technical problems and the costs, both in terms of expensive cylinders and the cost of compression, would seem to present considerable problems, but compression would expand the range of possible users for the gas and would allow for its sale.[11] If the gas can be sold then the number of people who might be willing to run and own biogas plants is increased by those who do not themselves have sufficient demand for the gas. If the gas can be transported, this would make possible the building of larger community-scale plants.

Problems of Consumer Surplus

The previous two sections have described a number of possible ways of valuing dung and gas. The actual valuation chosen will depend on a view of the real alternative use of the dung or the real substitute for the gas. The justification for the biogas plant is not only that there will be savings in alternative fuels (wood) and fertilizer (chemical nitrogen etc.), but that there will be a greater quantity of fuel and fertilizer available when the biogas plant starts operating. This greater quantity is accounted for automatically in the analysis with the estimation of the fuel and fertilizer output of the plant. However, the valuation suggested in the previous sections may have to be modified because it may not be legitimate to value all this extra fuel output in terms of the price ('cost') paid for the smaller quantity of fuel that was used prior to the introduction of biogas. The people cannot 'buy' the new increased quantity of gas at the old price.

Figure 24 shows the demand for fuel and two alternative supplies: wood and biogas. With the introduction of biogas the quantity of fuel used rises from q_1 to q_2. It has been suggested above that the valuation could either be in terms of the cost of wood p_1 or the cost of biogas production p_2. If p_1 is used this sets the gross benefit of the project equal to $p_1 \times q_2$ (or the rectangle $p_1\, r\, q_2\, o$). From this would be subtracted the previous situation $p_1 \times q_1$, to determine the net effects. This clearly overestimates the benefits of the scheme by the area *urs*. The comparison between $p_1 \times q_1$ with $p_2 \times q_2$ underestimates the benefits by the area *uts*.

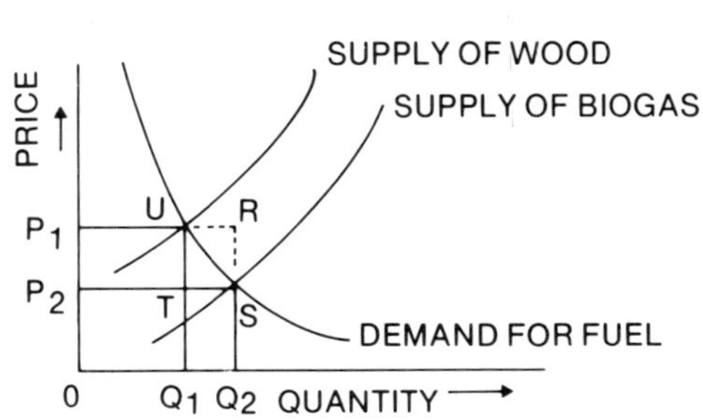

Fig. 24. Diagrammatic representation of the demand for fuel and the supply of wood and biogas.

[11]Both Prasad et al. (1974, p. 1358) and Makhijani (1976, p. 17) feel that the storage of large quantities of methane is prohibitively expensive under current village conditions.

This difference in valuation would only matter if the demand for fuel were, in fact, downward sloping as shown in Fig. 24. If in any real situation, over the relevant range of prices and quantities, the demand for fuel was not influenced by price (i.e. the demand curve in Fig. 24 was horizontal), then the valuation of gas in terms of the wood price would be correct. But if the demand curve does slope downward, the correct procedure is to measure the gross benefit to the project in terms of the whole area $p_1\, u\, s\, q_2$. This may be approximated to the average between the old and the new price and the average of the new and old quantity:

$$[(p_1 + p_2)/2] \times [q_1 + q_2)/2]$$

A corresponding problem arises with the valuation of the gas in terms of the labour cost of collecting an equivalent amount of wood. The labour saved as a result of using the gas cannot exceed the amount of labour actually used in the situation without the biogas plant when the wood was collected. It would be an error to attribute to the gas the value saved had a larger amount of wood (equivalent to the increased volume of fuel provided by the gas) been collected. This kind of error is often made when the cost of wood is calculated per unit weight (kg) and this unit cost figure is then used to value the increased amount of fuel provided by the gas.

Labour

The valuation of labour is given extensive treatment in project appraisal manuals because the employment of labour is often considered both a cost and a benefit to the project. In relation to biogas plants, the labour input to family-sized plants for mixing and charging is often quite small (of the order of 20 minutes per day), and it may well be that the use of this labour has no opportunity cost. However, the valuation of labour used for carrying and spreading the spent slurry will depend crucially on the characteristics of the labour situation in the village and on previous practices for handling fertilizer and dung.[12] The transportation of the wet slurry will be more difficult and costly than the transport of the much drier dung. If the slurry is dried, certain nutrients (in the water soluble ammonia) will be lost. In practice, it might be expected that at certain times of the year the slurry would be stored and dried and at others, put straight on to the fields. The valuation of labour can therefore be the deciding factor in the comparison of large- and small-scale techniques (see Disney 1976 and case studies presented later).

Labour is clearly not a homogeneous category. Some of the tasks associated with biogas can be done by unemployed unskilled labour (possibly the labour of children), but other tasks require considerably more skill. Each type of labour might be expected to have a different opportunity cost. The opportunity cost of women's labour is likely to be different from that of men in many societies.

In many developing countries the value of skilled labour may be underestimated by its market price. The value of a skilled technician needed to rectify the problems of a biogas plant may be much greater to the farmer than the cash amount that he is charged. It is important therefore to cost into the analysis of biogas a realistic valuation for skilled labour. At the limit, certain levels of skill just will not be available at the village level and the cost of such skills is effectively infinitely high — a plant design that requires these skills cannot work in the village situation. Such skills might be associated with the maintenances of gas compression equipment.

Capital

Capital is extensively treated in the cost-benefit manuals in terms of its opportunity cost, and in terms of problems of 'discounting' costs and benefits that occur over differing time profiles to a common 'present value' (Squire and van der Tak 1975, p. 75). In choices associated with village investment, what constitutes capital, its availability, and its alternative uses is difficult to establish.

[12]Berger (1976, p. 4) estimates that about an hour is necessary for diluting and mixing slurry. There are no estimates available for the amount of time involved in disposing of slurry.

If the capital referred to is at least partly the funds that the government is willing to invest in the scheme, then its opportunity cost might be expected to differ from the opportunity cost of the capital controlled by the people in the village.

Many of the designs for biogas that currently exist are very expensive in terms of the sorts of (capital) items that have to be purchased from outside the village — particularly the cost of the sheet steel that is often used for the gas storage container.[13] New designs are clearly possible that reduce the number of these purchased items either by switching to cheaper materials (wood and plastics) or by switching to materials that exist in the village and have very low opportunity costs (oil drums, local bricks, water).

With village investment, the absolute size of the 'initial investment' is an important factor. It is of little importance that the rate of return is good, or the net present value (NPV) is positive, if the necessary level of initial capital is not available to the villager. Access to capital is an essential part of the political process and capital is clearly not rationed solely on the basis of its cost (the rate of interest). Access to capital will be a major determinant in the rate of adoption of biogas plants and therefore in the pattern of their ownership.

Where loans are involved, the cash returns to the project also become important. Although many of the opportunity-cost benefits of the biogas investment may be quite large, the cash value of these returns might be insufficient to pay back required loans (Moulik and Srivastava 1975, p. 60-62). When this is the case, there may be good reason for the government to subsidize the loan so that the social returns rather than financial returns are maximized.[14] An analysis based solely on the cash transactions associated with the investment is therefore a necessary part of the evaluation of biogas systems.

Other Outputs

A number of other outputs associated with biogas investments, which pose particular problems of quantification and valuation, were dealt with in general terms earlier. An output such as the reduction in the transmission of human pathogens, which could result from the processing of human waste through a biogas plant, is a particular case in point. The increase in the value of human waste when it can be gasified might help encourage people to dispose of excreta safely. It is unlikely to be possible to specify in precise quantitative terms the reduction in illness that would result from the safe disposal of excreta because many other factors influence health. The safe disposal of excreta may be a necessary condition for the improvement of health, but it is not a sufficient condition. Even if the effect on health could be quantified there would be other problems in placing values on such a reduction.

This problem can be approached in two ways. First, when biogas is being compared with other 'high' levels of technology, such as the factory production of urea, the benefits from the safe disposal of human waste can be introduced as an explicit, but qualitative, objective in the decision matrix discussed earlier. It is unlikely that the factory production of urea will have any public health benefit in the villages, and it is then possible to see what value would have to be placed on the safe disposal of excreta to make the biogas plant as 'profitable' as the urea plant (it is, of course, possible that no such weighting is necessary).

Second, when the comparison is among different biogas designs, all of which have some potential public health benefits, a conflict can arise between the need to dispose of human waste safely and the objective of producing the maximum amount of gas per day in a digester of fixed size. This can occur when the retention time

---

[13]Berger (1976, p. 13) shows the cost of the cover is over one-third of the total construction costs. Prasad et al. (1974) give 35% quoting KVIC (1975, p. 1355).

[14]In Korea when the government terminated its heavy subsidy to the construction costs of biogas plants there was nearly a complete cessation in the number of plants installed.

needed to kill all the pathogens in the excreta exceeds the retention time that would give the maximum gas production per unit of time. In this case, it is only possible to show the reduction in the daily production of gas that would result from extending the retention time long enough to kill the pathogens. It is then left to the decision-maker to decide whether the benefit of improved health is worth the cost of the gas lost.

Benefits resulting from the possible improvement in the domestic environment following the reduction in wood and dung smoke that results from the introduction of biogas would have to be accounted for in a similar way.

The reduction in the time taken for cooking when using biogas compared with using dung and wood would first be balanced against the extra time taken in managing the biogas plant. Any surplus labour would then be valued according to the opportunity costs of that particular type of labour.

Opportunity Costs in Relation to Chains or Systems of Investment

Biogas plants are often advocated as part of a larger system in which the output of one part of the system becomes an input to another. There is evidence to suggest that additional investments to provide inputs or to use outputs can raise the return from the biogas plant (for instance, when the liquid from the slurry of biogas plants is connected to ponds in which algae are grown and subsequently fed to fish etc. — see chapter by Pyle). This poses no problem to the opportunity-cost analysis (all investments are in principle part of such a system), but in practice, care has to be taken in sorting the net effects on the whole system from these additional upstream and downstream investments. The benefits to the biogas investment from putting part of the slurry through ponds to grow algae and then fish, will be the value of the fish minus the cost of constructing and running the various ponds. Investment in these projects would have to be compared with the value of the liquid part of the slurry in its alternative use — most likely the returns from putting it on the fields and the subsequent increase in crop yields. The advantages of taking the fish production option might be savings in transport costs of the wet slurry to the fields and the greater utilization by the algae of the plant nutrients in the slurry.

It is not always necessary to evaluate the various intermediate outputs (such as the slurry) in the chain because the purpose of the analysis is to weigh the primary inputs to the system against the final outputs. This valuation can only take place when the system has reached a stable state. There is a certain danger of 'double counting' in the analysis of systems when intermediate products are valued: first, any commodity must be ascribed the same value regardless of whether it is an output or an input of the system; second, the value of an intermediate output cannot be valued as a net benefit to the project if it is subsequently used as an input elsewhere in the system. It would seem that in one study this mistake was made and the values of the slurry (as fertilizer), the algae (as protein), and the fish (as protein) were all counted as net benefits to the same project (Philippines de la Salla University).

Perhaps the largest source of error in the evaluation of these chains of investment is that the physical input/output relations of the system are incorrectly estimated. The proponents of these systems become over-enthusiastic and leave out of the analysis certain inputs that are required to make the cycle work. Examples can be found of systems that seem to be out of balance (Tyler 1973).

Case Studies

The following case studies have been selected to illustrate different approaches to the evaluation of biogas technology and to highlight some of the points made previously. The number of thorough economic evaluations of biogas is very small and those chosen here are among the best, but the purpose of this section is to point out possible points of weakness in the case studies rather than to praise them. The examination of these case studies concentrates on the conceptual problems involved rather than on an evaluation of the empirical data. It is important to note that quite wide variations in many of the crucial parameters are to be found among the case studies, even when similar plants are involved, and this raises some doubts as to the precise conclusions that can be drawn from these studies (see for instance Table 20).

Case Study 1

(ICAR 1976, particularly pages 25-35, 61-62)

This is a very professional report that attempts to fill the gap in detailed information about the viability of biogas plants in India. A comparison is made of six sizes of biogas plant and current fuel and fertilizer practices. No consideration is given to the wider aspects of other village investments or to alternative means of satisfying fuel and fertilizer needs. It is stated (on p. 25-26) that the net returns per plant per year (R) from the investment in each size of biogas plant, over the returns from the existing practice for fuel and fertilizer, is equal to the gross value of the methane gas and the slurry (A), minus the value of the dung, had part of it been burned directly and the rest used as farm yard manure (B), minus the cost of the maintenance of the biogas plant (E). So that R = (A - B - E). The net present value is then the discounted value of R, minus the investment cost (I).

The gas is valued according to the market price of a thermally equivalent amount of kerosene; the slurry is valued as a fertilizer, but no indication is given as to the method of valuation; and the dung is valued both in terms of its fertilizer value (but again no method is given) and the market price of its equivalent thermal value of kerosene. This analysis illustrates how simple 'sensitivity' analysis can be carried out to show how sensitive the profitability of the schemes are to changes in the value of: the efficiency with which gas is produced; the proportion of dung that was burned in the previous situation; and different ways of valuing the methane gas.

By way of illustration, the numbers are presented for a 60 ft³/day (1.7m³/day) plant using the assumptions that there is a 50% efficiency of gas production, the gas is valued at the cost of a thermally equivalent amount of kerosene, and either all the dung was previously burned (*a*) or all the dung was previously used as fertilizer (*b*).

Inputs per year
Dung + Generator + Labour

Outputs per year
Gas + Slurry fertilizer

Each of these items can be valued according to the data supplied by ICAR on an annual basis.

The opportunity cost of the dung:

(*a*) either as fuel $P_2 \times D$
$= 0.0545 \times 730 \times 5$
= Rs198.92/year; or

(*b*) as fertilizer $P_3 \times M_1$
$= 0.04 \times 2.50 \times 730 \times 5$
= Rs365/year

(*c*) the opportunity cost of the generator, including the labour cost of running it:
$P_{1(g)} \times 60$ ft³ × (1-% downtime) × 365
$= 0.0165 \times 60 \times 0.9 \times 365$
= Rs329

(*d*) the opportunity cost of the gas:
$P_{1(t)} \times C \times K \times 365$
$= 0.0186 \times 0.5 \times 60 \times 365$
= RS203.67

(*e*) the opportunity cost of the slurry as fertilizer:
$P_4 \times M_2$
$= 0.05 \times 3.65 \times 730 \times 5$
= Rs666.125

Where:

K= capacity of the cow dung gas plant (ft³/day) (1 ft³ = 0.028 m³)
I= investment in the gas plant (Rs)
d= dung produced per animal per day (kg)
N= number of animals required per plant
W= wet dung processed per plant per year = 365 Nd(kg)

D= dry dung available per year from dung of equivalent quantity as processed in a gas plant (kg)
G= methane gas produced per plant per year = 365 K (ft^3)
M_1= farmyard manure obtained per year from dung of equivalent quantity as processed in a gas plant (kg)
M_2= gobar-gas manure obtained per plant per year (kg)
c= efficiency of gas production (varying from 0 to 1)
f= proportion of cow dung used as fuel in the existing system (varying from 0 to 1)
1-f= proportion of cow dung used for making manure in the existing system
A= gross return per year from cow dung gas plant from value of methane gas and gobar-gas manure (Rs)
B= gross return from existing practice from value of dung fuel and farmyard manure (Rs)
E= recurring maintenance expenditure per plant per year (Rs)
R= net return per plant per year over existing usage from the investment in a gas plant (Rs)
= $(A - B - E)$
i= rate of interest (per rupee per annum)
n= economic life of the plant in years
P_1= price of methane gas in Rs per ft^3 (subscript g is generation cost, t is thermal equivalent cost)
P_2= price of dung fuel in Rs per kg
P_3= price of farmyard manure in Rs per kg
P_4= price of gobar-gas manure in Rs per kg

Therefore:

	Inputs per year	Outputs per year
(*a*)	Rs 198.92 + 329	Rs 666.125 + 203.6
(*b*)	Rs 365 + 329	Rs 666.125 + 203.6

Therefore profit, per year = (*a*) Rs342 and (*b*) Rs176.

The net present value (NPV) is given by the formula $(1-(1+r)^{-n})/r$, and for a project life of 20 years and at 10% interest it equals 8.5136 (where r = 0.1 and n = 2). Therefore, NPV = (*a*) 2912 and (*b*) 1498. This compares with results in the ICAR report that show a = 3460 and b = 2040. The difference is due to the exclusion of the labour costs of running the plant in the ICAR calculation.

A number of points emerge from this study.

It is worth noting that the above calculation is equivalent to comparing the situation with the investment and the situation without it; this can be easily shown:

A. The situation 'with' is
Dung + Generator + Labour ⟶
Gas Heat + Fertilizer
B. The situation 'without' is
Dung + Stove + Labour ⟶ Dung Heat

If B is subtracted from A, and the stove is considered a 'sunk cost' because it is an expense incurred before the decision to go ahead with biogas, the net effect is:

Generator → Gas Heat - Dung Heat + Fertilizer

which is the same as:

Dung (Heat) + Generator → Gas(Heat) + Slurry Fertilizer

The value of the slurry in this calculation is far greater than the value of the gas (Rs666 compared with Rs204/year). But it is unnecessary to anaerobically digest dung unless gas is required. It would certainly be worth comparing the situation of composting-plus-kerosene with the biogas situation; the gross returns of the compost might be lower, but it might involve considerably less investment than biogas.

The slurry was considered to be much more valuable than the dung by ICAR. This is possible, but again the opportunity cost of the dung should perhaps be the net value of composting the dung rather than its value unprocessed in any way.

Certain costs were not apparently included in the analysis: water; cooking equipment for use with gas; costs of distributing the gas (though this is mentioned); and the transport of kerosene (and fertilizer?).

No explanation is given for the valuation of dung and slurry, nor is this considered sufficiently important to merit inclusion in the sensitivity analysis. Inputs (such as crop residues) other than dung are also not considered in the calculations, though they are mentioned.

In the valuation of gas in terms of kerosene, there is no discussion of the use of market prices. For instance, a shadow foreign exchange rate might have been used to more truly reflect the cost to the economy of kerosene. The gas might also have been valued in terms of the thermal equivalent of wood; wood is mentioned as an expensive fuel in the report.

The sensitivity of the calculations to changes in the life of the project and the interest rate were calculated. Reductions in interest rate and increases in life expectancy have the effect of improving the net present values.

The problems of income distribution were considered to the extent that it is stated that: "it would be evident from the above description that the present plant designs are beneficial solely to the cattle-owning rural rich."

The private cost-benefit analysis is certainly an attempt to bring out the important point that the returns to the actual investor are somewhat different from the social returns. It is not clear, however, why it is assumed that the farmer would have paid cash for chemical fertilizers had he not had the biogas plant; nor is it clear why the farmer would not also have paid cash for kerosene. The results would also have been very different if the fertilizer value of the dung had been raised by a composting process.

Case Study 2

(Disney 1976, particularly pages 9-15)

The paper starts with the view that "the superficial attractiveness of intermediate technology may . . . be overstated by some of its more enthusiastic supporters." To support this view a comparison is made of the production of nitrogen fertilizer by existing designs of biogas plants (in India) and by the more conventional process that produces chemical nitrogen in the form of urea. Calculations are carried out to show the cost per tonne of nitrogen produced by each technology. If the capital costs associated with the production of an equivalent amount of nitrogen are correct, current biogas plants use between 2.5 and 8 times more capital per unit of nitrogen output than conventional plants. These intermediate techniques are not necessarily less capital intensive (per unit of output) than the 'high' technology, but as Disney also mentions, biogas plants not only produce nitrogen but also gas. It is argued in his paper that with the historical process of technical change in which costs have tended to be reduced largely at the capital intensive (K/L) end of the spectrum of available techniques, the superiority of western technology is to be expected.

The study is well argued and illustrates another form of the evaluation of biogas — this time in comparison with another provider of fertilizer. The gas from the biogas plant is treated as a 'negative cost' (i.e. as a benefit) to the production of nitrogen. The costs of transporting chemical nitrogen to the user, which are essential in a comparison of this kind, are included in the analysis.

A partial sensitivity analysis was carried out for changes in both the shadow price for foreign exchange and the cost of transporting the urea for the conventional plant. For the biogas plant, sensitivity to changes in the capital costs, the proportion of nitrogen in the slurry, and the opportunity cost of the dung were considered.

Gas is valued in terms of a thermally equivalent amount of electricity (at Rs0.012/ft^3; 1 ft^3 = 0.028m^3); the cost of dung is allowed to vary widely between Rs100 and 20 per tonne and is based to some extent on coal as an alternative fuel; and the slurry is valued in terms of its nitrogen content (1.6%).

The major problems with this analysis are that labour is inadequately valued for the biogas plants, and the selection of variables to be altered in the sensitivity analysis might be considered biased. (Some of these problems have been corrected in the revised version published in "Development and Change.")

Labour is represented as 86% of the total costs of the small biogas plant's costs and 37% of the larger plant's costs (assuming the 'high' cost of dung). The valuation of labour is therefore crucial in determining the competitiveness of the biogas system. A figure of Rs12/day is used even though the Indian Council for Agricultural Research used Rs4/day. Disney does adjust his labour costs to some extent before making his final calculations by reducing the amount of manpower required to run the plants, but the labour bill is still a substantial input. In certain village situations it might be expected that labour would have a very low (even zero) opportunity cost and some people consider that one of the greatest advantages of the biogas system is that it can utilize this resource. If a zero opportunity cost of labour is used to test the sensitivity of the analysis to changes in the cost of labour, all the biogas plants produce nitrogen more cheaply than the smallest conventional plant (with the exception of the smallest biogas plant when the dung price is 'high', see Table 34).

The sensitivity of the paper's conclusions might also have been tested against changes in other variables.

The ratio of slurry to gas (27 kg/150 ft^3) is similar to the figures used by Prasad et al. 1974, but less than the figures used by the Indian Council for Agricultural Research (which would be equivalent to 87 kg at the 150 ft^3 level) and by Sathianathan (1975) who uses a range of between 27 and 99 kg at 150 ft^3. These increases would considerably influence the balance of costs and reduce the cost per unit of nitrogen produced.

The value of gas (Rs0.012) is less than the figure used by ICAR (Rs0.0186). A value such as ICAR's produces a gas value per tonne of nitrogen of Rs6500 rather than the Rs4200 used by Disney. If this lower figure is used the cost per tonne of nitrogen is further reduced by (6500-4200) Rs2300, making the larger biogas plants cheaper than all the conventional urea techniques even using the 'high' cost of dung. It is not clear why the gas should have been valued in terms of electricity rather than wood, kerosene, or even coal as coal was used to establish the opportunity cost of dung.

The volume of gas actually produced is assumed equal to the plant's design capacity. This is unlikely and therefore the cost per tonne of nitrogen would be higher than is suggested here.

Two further points are of interest.

No interest is charged for the use of the capital in any of the calculations in this paper. It is usual to calculate annual capital costs as an 'annuity' — what equal annual amounts have a present value equal to the cost of the capital: for example

$$\text{Annual Payment} = K \div \left(\frac{1-(1+r)^{-n}}{r}\right)$$

where K = capital = 3360, r = 10% per annum, and n = 10.

The value of the bracket can be found in discounting tables (e.g. Lawson and Windle 1974): 3360÷6.1446 = 546 annually. This is more correct than taking the total interest payable over the 10 years (3360 × 0.1 × 10 = 3360) and adding this to the capital costs (3360 + 3360 = 6720) and spreading this equally over ten years (6720 ÷ 10 = 672 annually).

An essential assumption of Disney's analysis is that slurry has no value other than as nitrogen. This is not in fact the case because the slurry also contains humus and

Table 34. Costs (Rs) per tonne of producing nitrogen using biogas plants, when the cost of dung is assumed to be either 'high' or 'low,' compared with urea plants.

Cost of dung	Small biogas plant	Large biogas plant	Small urea plant	Large urea plant
High	4618.75	2109.38	2332.00	1828.70
Low	-681.25	-2034.37		

trace elements. Therefore, the value of these extra elements should be subtracted from the cost of producing nitrogen in the biogas system.

Case Study 3

(Prasad et al. 1974, particularly pages 1353, 1355, 1364)

This is perhaps the most widely quoted and influential article written on biogas in recent years. It is a masterly marshaling of a huge amount of data into a coherent and easily read argument. It raises a number of issues and considers many of the alternatives with which biogas can be compared. The comparisons between biogas and rural electrification on the one hand, and biogas and urea plants on the other, represent only a small proportion of the paper and are "intended mainly to stimulate detailed cost-benefit analysis of alternatives." However, for the purposes of this case study only these two comparisons are considered.

Comparison with Electricity

A quite wide range of inputs and outputs are identified for a 5000 ft^3/day (140 m^3/day) plant and some of them are quantified. The largest omission from the list is the cost (the opportunity cost) of the dung. The assumption implicit in an analysis presented in this form is that the dung has no other use (a zero opportunity cost) and that only the cost associated with its use is the cost of collection. The inclusion of an opportunity cost for the dung would make quite a difference to the apparent viability of the biogas plants. If the figures used by the Indian Council for Agricultural Research are used (and these seem to be on the high side) the cost of the necessary dung would have been in the region of Rs13412/year (if the dung had previously been burned) and Rs24609/year (had the dung been used as a fertilizer). This is compared with a benefit stream of Rs38300/year and a running cost (excluding capital) of Rs12206. If the ICAR figures are used then a corresponding adjustment would have to be made to increase the value of the slurry.

The point that is being made here is not whether any one particular value is correct but rather that it is unlikely that the fertilizer and the gas benefits could be obtained without some (opportunity) costs elsewhere associated with the use of the dung, etc.

The estimation of the annual capital costs of the plant is omitted from early calculations, but is included in more detailed calculation at Rs4350 per year. This is an annuity equivalent to a 30-year plant life and an interest rate of 10%/year. If these costs are included, the total annual costs of the 5000-ft^3 plant rise by 36% from Rs12206 to 16556.

If the figures used for rural electrification are correct, it appears that there is a massive subsidy being paid to electricity generation. This could either be a direct subsidy from the government or it could be an indirect subsidy involving the government underwriting the losses of the electricity company that these figures suggest. A third possibility is that the subsidy is a transfer payment from the urban population if the profits arising from electricity sales in urban areas are sufficient to allow the electricity company to sell electricity to the rural population at the figures quoted.

The comparison of biogas with electricity also illustrates a possible confusion between 'prices' and 'values.' For instance, no value is attached to the free provision of gas for cooking to all the village houses, but at the same time, prices are attached to the slurry that is produced by the biogas plant, but which may well not actually be bought for cash by the villagers. It is essential to separate the analysis into two distinct parts: one in which both the costs and the benefits are valued according to their 'social opportunity cost' and another in which the costs and benefits are valued according to the actual cash transactions that occur.

Comparison with a Urea Plant

This is a considerably more rudimentary analysis than the comparison with electricity. It attempts merely to show the rough orders of magnitude associated with each system. The comparison is made largely on the basis of the ratios of capital to output (or capital to turnover) and capital to labour. This formulation makes it difficult to compare the two processes as the costs per tonne of nitrogen delivered to the farmer cannot be calculated because costs are not given for the inputs to each process: dung, naphtha, labour, and the transport of the urea. Again, there is a possible implication that the inputs, particularly of dung and other compostable material, are considered to have a zero cost.

It is interesting to note that it is assumed here that a plant that produces about 5000 ft^3 of gas per day produces about 8.8 tonnes of nitrogen per year. This is considerably higher than the figure used by Disney (5.8 tonnes/year from a 5600-ft^3 plant) and is higher than the figures that appear elsewhere in the paper: 2.7 tonnes in the urea calculation and 4.4. tonnes in the electricity calculation. The large amount of nitrogen contained in the slurry is partly due to the fact that it is composted with "refuse etc." If this is the case then it might be expected that some value for the refuse should be added to the costs side of the calculation. It is not clear why composting is thought likely to only occur in conjunction with biogas production.

Case Study 4

(Moulik and Srivastava 1975)

This is a thorough piece of research that concentrates on the State of Gujarat where about 28% of all Indian biogas plants are located. The study details the location, ownership pattern, and running problems of biogas plants in the State and discusses a number of the important issues in the operation of the biogas scheme. The economic aspects of biogas are covered and an attempt is made to answer the question: "If farmers are rational and the economics are as good as claimed by the KVIC experts, why does the demand for biogas plants not increase rapidly?"

Economic and financial analyses are carried out on nine plant designs ranging in size from 60 to 1250 ft^3/day (1.7 to 35 m^3/day). The data used are largely from the Khadi Village Industries Commission and might well be thought to be overly optimistic. This problem with the data is recognized by the authors and some attempt is made to adjust the KVIC figures on the basis of survey data. The main adjustment is in the amount of gas that is likely to be produced from the plants per year, because it was found that the actual output was less than the design capacity suggested by the KVIC.

A number of costs and benefits are listed. Explicit account is taken of the costs of the ancillary investment associated with pipes and appliances and with the costs of keeping the gas-holder well painted. The gas is valued in terms of the market cost of a thermally equivalent amount of kerosene. The dung and slurry are given money values based on data supplied by the KVIC, but no indication is given of the basis on which the figures were calculated (the slurry is said to be 1.875 times more valuable than the equivalent amount of dung). Different life spans were assumed for various parts of the plant: the civil construction was assumed to last 40 years; the gas holder 10 years; and the pipes etc. 30 years. Various indicators of investment worth are calculated on the basis of interest rates of 10, 13, and 15%. The only plant to have a negative net present value was the 60-ft^3 plant when the interest rate was 15%.

A number of points are of interest.

No mention is made in the analysis of the costs associated with the labour used to run the plant, nor is any cost attributed to the water used. The inclusion of these two costs would have reduced the practicality of the plants.

Just over 40% of the benefits of the scheme arise from the difference in the values of the dung and slurry.

The method used for the valuation of the dung and slurry is not made explicit and it is therefore not clear how the opportunity costs are derived. It is, however, suggested that agricultural wastes are the normal fuel in the area and that these have negligible costs to the farmer. If this were the case, then the benefits of the biogas plant would be the differential value between dung and slurry as a fertilizer (and questions might be asked about methods of improving the value of the dung through aerobic fermentation), the value of the saving of the agricultural wastes (which might be small), and the value of the gas. The gas might be valued in terms of the heat equivalent of the agricultural wastes, or it might be valued in terms of a thermally equivalent amount of kerosene. But, if kerosene were used, this would raise the question of whether the agricultural waste might also have been valued according to its opportunity cost in terms of kerosene (an adjustment might be made to take into account the fact that biogas can be used for lighting and for driving engines whereas agricultural wastes cannot). The problem can be shown simply:

Without biogas
Inputs: dung (as fertilizer) + agricultural wastes (as fertilizer). Outputs: dung (as fertilizer) + agricultural wastes (as heat).

With biogas
Inputs: dung (as fertilizer). Outputs: slurry (as fertilizer) + gas (heat).

By subtracting the 'without biogas' situation from the situation 'with biogas' the net effect is: Inputs: dung (as fertilizer) + agricultural waste (as fertilizer). Outputs: slurry (as fertilizer) + gas (as heat) - agricultural waste (as heat). This is equivalent to saying that the benefits of gas, slurry, and agricultural waste (as fertilizer) are gained for the loss of the dung as fertilizer and the loss of heat from the agricultural waste.

The financial analysis differs only from the economic analysis in the subsidy to the farmer for the purchase of the biogas plant. This would again appear to illustrate a possible confusion between the analysis of the project from the point of view of opportunity costs and from the point of view of cash transactions. The financial analysis does bring out the point that the actual returns as perceived by the farmer can be quite small and that this could restrict the rate of acceptance of a project that is 'socially' beneficial.

This is the only study that explicitly compares the returns to biogas with the returns that can be achieved by other rural investments such as buffalo, tractors, and lift irrigation. However, it is not stated whether the analyses were carried out on a comparable basis.

The fact that very little of the data used in the analysis was obtained from actual operating situations weakens the conclusions that are based on it. This is a useful 'back of the envelope' calculation, but it could not be considered as an authoritative statement as to the viability of biogas in India.

Case Study 5

(Berger 1976)

This is an extremely thorough analysis of a 100-ft^3/day (2.8-m^3/day) plant under Nepalese conditions. It involves a complete statement of costs and benefits (though water is excluded) that includes a more than usually detailed discussion of the possible valuations that can be placed on gas, dung, and slurry. The possible values for the gas from a 100-ft^3 plant range from Rs836/year (valued in terms of charcoal) to Rs5055 when valued in terms of gasoline. The annual value of the gas in terms of electricity is Rs2263. The value of the fertilizer from the plant is taken as the nitrogen value of the slurry minus the nitrogen value of the dung (i.e. the dung is assumed to have been burned in the situation without the biogas plant). A variety of other assumptions are discussed that show the variation in the nitrogen value of the dung or slurry depending on the way the dung/slurry is used in the fields.

A simplified method is used for the treatment of capital: the plant is depreciated over 25 years in equal annual amounts and the interest is taken as the interest on half the capital value per year (i.e. 5000 divided by 2, times 15% equals 375 per year). This is a commonly used approximation, but it does produce an annual capital cost that is substantially less than the amount that is produced using the annuity method described previously. The annuity method produces an annual cost of Rs773 per year for 25 years at 15% with no scrap value.

The net result of the analysis shows that annual benefits exceed annual cost by

Rs5545. These would have been reduced to Rs347 had the annuity method been used for the valuation of capital.

The analysis assumes that the cow dung would previously have been burned, though the author points out that those who might consider using the biogas plants are currently burning wood. Furthermore, it is assumed that 100 ft^3 of gas would actually be produced per day — this might be considered far too optimistic. But the analysis is set out so clearly that it would not be difficult to do as the author suggests — "to the extent that individual farm conditions are different from the assumption the analysis can be modified." The only slight problem with the chosen format is that it is not obvious at first sight that the benefits from the slurry are net of the costs of the nitrogen that could have been obtained merely from using the dung directly on the fields.

The Social and Economic Determinants of the Demand for Biogas

The answer to a number of important macroquestions cannot be supplied by the kind of microanalysis discussed so far. In particular, the introduction of biogas on a wide scale has implications for macroplanning such as the allocation of government investment and the effects on the balance of payments etc.[15] Furthermore, many of the factors that determine the rate of acceptance of biogas plants, such as credit facilities and technical backup services, are likely to have to be planned as part of a general macropolicy. The importance of biogas from the point of view of the allocation of research and development funds also requires a more macroview of the technology.

Although the techniques for the production of biogas are likely to undergo substantial change in the coming years (particularly reduction of capital costs and ease of operation) it is still possible to broadly describe the social and physical characteristics of rural areas in which biogas plants are most likely to be viable (or are least likely to fail). The prevalence of these characteristics will give a more macroindication of the possible importance of biogas in the country's rural development, fuel, and fertilizer policies.

From the framework provided by social cost-benefit analysis it can be assumed that biogas will be most viable in those situations in which the necessary inputs have a low opportunity cost, where the efficiency of the operation of the plant is 'adequate,' and where the alternatives to the outputs from biogas plants have a high opportunity cost. Surveys would have to be carried out to determine the physical location of those areas that contain these characteristics.

Necessary Inputs have a Low Opportunity Cost

This is most likely to occur where: (1) agriculture is such that sufficient amounts of material from which methane gas can be produced are available with an opportunity cost that is at least no higher than when it is used as a fertilizer or fuel; (2) industries exist (e.g. paper production, distilleries) that produce, as by-products, large amounts of material from which methane can be produced; (3) and/or there is no social restriction on the use of human waste; (4) and/or cow dung is traditionally collected; and (5) water is available and can easily be fed into the digester; (6) capital is available with a sufficiently low opportunity cost as a result of either the adequate supply of capital or because of the lack of opportunities for its alternative use — it is to be expected that methane generation will only become an attractive opportunity when

[15]Makhijani and Poole (1975) suggest that current biogas designs would use up 20% of India's annual production of steel and 25% of her cement production if biogas plants were applied to 10 million hectares of cultivatable land per year (p. 88). Mardon (1976) shows (p. 9) that at the planned rate of biogas implementation of 100 000 by 1978, it would take 300 years to satisfy the energy needs of the 5% of the population who own 5-6 cattle in India.

certain other investments, such as irrigation, have been carried out; and (7) labour is available and willing to undertake the work of operating the methane generator on a continuous basis.

Efficiency of the Operation of the Plant is Adequate[16]

This is most likely to be possible where: (1) there is uniform input of gas producing material (the plants seem to be simplest to run when only one type of input, such as cow dung, is used); (2) technical advice and technical skills are available both for the construction of the plant and in trouble-shooting — this might be most likely to occur near towns; (3) smaller size plants can be avoided — there is less problem of shock loading with larger plants and with larger plants it becomes possible to employ and train full-time operators; (4) the surrounding air temperature does not fall below the level at which methanogenic bacteria operate adequately (above 15 °C for the bacteria commonly found in generators); (5) plant design is adequate — e.g. no places in which blockages can occur or water condense; and (6) maintenance and good operating practice is likely to be carried out.

Alternatives to the Outputs from Biogas Plants have a High Opportunity Cost

This is likely to occur where: (1) there is a physical limit to the amount of fuel and fertilizer available from an alternative source — this might occur as a result of the cost of transport or from policy decisions as to the rate and geographical direction of expansion in the larger scale fuel and fertilizer industries[17]; (2) and/or there is a scarcity of wood; (3) and/or dung is being burned as a fuel; and (4) there is insufficient water to make use of chemical fertilizers; (5) there is insufficient cash to purchase other fuels and fertilizers; (6) the use for the gas is near the generator (or simple and economic compression facilities exist); and (7) the cost of handling the slurry is not large enough to reduce the net value of the slurry to unacceptable levels (some plant locations make slurry handling particularly difficult and expensive).

Some of these characteristics can be modified by government policy and expenditure, but it should be relatively easy to identify the relevant area, particularly on the basis of the third set of characteristics — where alternatives to biogas and slurry are costly. Which areas are isolated from alternatives by high transport costs? Which areas are experiencing unacceptable levels of depletion of forest resources? Where is dung currently burned? Answers to such questions would be available from a survey of rural energy needs and current fuel and fertilizer practices.

The areas in which biogas might be most viable can be further narrowed down by considering the factors in the second group that affect running efficiency. Areas with low night or winter temperatures can be excluded; areas can be identified in which technical advice is available or in which the government is prepared to make it available through rural development policies or policies to help certain groups; and locations (such as intensive animal rearing units) can be identified that are known to have large amounts of uniform outputs from which gas can be generated.

The availability of inputs, within the areas identified so far, will be determined more by the social groups concerned than by purely geographical characteristics. Any number of social groups can be considered, but eight groups are suggested by way of illustration: agricultural or other business, with intensive animal or crop production; cooperatives formed to produce biogas; large existing

[16]Finlay (1976) is an excellent trouble-shooting guide to the problems encountered in the construction and maintenance of biogas plants.

[17]OECD (1968, p. 59) — a significant percentage of farmers surveyed for this report state that lack of access to fertilizer outlets was the main reason for not using chemical fertilizer.

social groups that are able to cooperate — such as large families, communal cooking groups, and cooperatives formed for other purposes; large farmers; small farmers; landless labourers; traditional collectors of cow dung; and women.

Unless the output from the biogas plants can be sold or bartered, investment in biogas will only be viable to those groups who have both access to the various inputs and have sufficient use for the outputs. This link between inputs and the need for the output is particularly important with the gas itself. If there is no way of the plant owner profitably using the gas, then it might be better to compost the inputs rather than go to the trouble of producing methane. A number of studies assume that the gas can be compressed and transported in cylinders at reasonable cost (Prasad et al. 1974, p. 1358; Makhijani 1976 p. 17; Philippines da la Salla University). But, compression, whether into cylinders or through an extensive system of pipes, involves a 'higher' level of technical 'know-how' than the rest of the biogas system. The compression of methane will therefore not be a practical proposition in many developing countries.

Until the problems of gas sale and delivery have been overcome, it is likely that biogas investment will have to be concentrated on the first four of the eight groups described previously. Only they are able to bring together the necessary inputs and are able to gain the benefits from the gas either through cooking or through the use of machines (pumps for irrigation etc.). But, as has been pointed out in a number of studies of India (Prasad et al. 1974; Government of India (ICAR) 1976; Mardon 1976) the distribution of income is such that currently only the relatively well-off can afford biogas investments. This means that only a small proportion of the village's energy needs can be met in this way, and that there is a likelihood that dung and other wastes will be denied those people who traditionally collected them. If income distribution considerations are part of the government's objectives, then a strategy of introducing biogas plants (or any other rural technology) only to the richest groups may well be unacceptable. The remaining option is to encourage the use of larger (community scale) plants on some more cooperative basis. But, as the ICAR report puts it: "At present not a single community biogas plant exists in the country. The concept will remain just a theoretical slogan unless the attendant socio-economic and technical problems are investigated in some pilot plants in the public sector, located in different agro-climatic zones in the country." (Government of India (ICAR) 1976, p. 42; see also Moulik and Srivastava 1975, p. 50).

If sale of the gas is possible, a wider range of social groups could be involved in the generation of methane, but these other groups (such as the last four groups in the previous list) would find it difficult to assemble the other necessary inputs without some form of cooperation and assistance. For instance, where dung is currently burned, the introduction of biogas will require the most behaviour changes in activities normally undertaken by women, such as cooking and the making of dung cakes, but women as a group are likely to lack access to capital. An interesting suggestion in this context is that loans be made directly to these women to give them an interest in and control over such changes, and to provide a means of directly raising their economic and social status (personal communication, Prof Scarlet Epstein, Institute of Development Studies).

The Government's own policies, then, are likely to be the largest factor in determining the distribution of biogas plants within the areas narrowed down according to the characteristics described previously. The provision of capital is one such policy variable. But the extent to which the lack of capital will be a constraint to the future adoption of biogas plants cannot be inferred from the current situation because past practice involves both much higher capital costs per plant than are likely to apply in future, and because only the family-sized plants have been promoted. The Gujarat Study (Moulik and Srivastava 1975) has a number of useful suggestions about possible methods of finance in India, but the problems of rural credit are much wider than the problems that particularly apply in the financing of biogas plants and they are not the subject of this study. A second major area of government policy that could influence the rate of acceptance of biogas plants concerns the provision of technical backup services to the plant. Again, it is difficult to specify the characteristics of this backup, but considerable experience has now been gained with the working of agriculture extension agents and the lessons learned might be used to develop a model for the provision of technical 'extension' services.

An Approach to Research Priorities

Two ideas are central to the arguments in this paper: first, research into biogas techniques must be placed in the wider context of policies for rural development and rural energy needs; and second, research and development of a particular set of techniques must not be abstracted from the social and political context in which the techniques are to be applied. It follows from this that before starting on research on biogas, governments (and research organizations) must first decide how important biogas is likely to be in meeting the needs of rural peoples. At the most general level this involves establishing the range of other possible areas of village life that might also benefit from research and innovation; more specifically it requires an examination of rural energy needs, current practices for satisfying these needs, and alternative means of satisfying them in the future.

If biogas does appear initially to offer advantages over other alternatives, research and development must be directed toward meeting a specific set of objectives. In particular, the objective of maximizing the production of methane gas must be set against the objective of improving the distribution of income (and therefore energy) and the need to further the interests of particular groups within society. This trade-off will determine whether research is to concentrate on those plants that would be most easily accepted (which may be the plant that meets the needs of the rich farmer group) or on those plants that are likely to satisfy the energy needs of a wider range of social groups (when the community plant might be favoured).

The appropriateness of a particular package of biogas technology will vary from location to location depending on the objectives chosen and on the availability of resources. This suggests that research priorities will have to be specified in terms of a process for conducting the research rather than in terms of individual pieces of technical research. It would seem appropriate in these circumstances to develop a process in which the views of villagers and the views of the researchers are combined in the development of research activities. This will involve an iterative process in which, in principle, villagers specify their needs, technologists specify the technical options available, and the villagers restate their needs in relation to the options. An implication of this would be that the location of the research would largely be in specific rural areas rather than in the laboratories of urban research institutes.

This does not mean that sensible lists of technical research cannot already be drawn up (Moulik and Srivastava 1975, p. 53-58; Sathianathan 1975, p. 164-172; Government of India (ICAR) 1976, p. 41-42), but it does mean that much of this work will only be useful if the views of the potenial users are given an important role in the development of the technology. Three broad areas of economic and technical research already stand out.

Cost Reduction

The current designs for biogas production at the village level appear to be considerably more expensive than they need be. This high cost not only reduces the numbers of plants operating but it also reduces the advantages that village-level biogas plants have over larger-scale technologies (Prasad et al. 1974, p. 1355; Disney 1976). In particular, many current designs involve very much higher amounts of capital per unit of output than more conventional means of supplying energy (Disney 1976, p. 9).

Cost reductions can be achieved through the use of other materials (plastics or locally available wood, water, and brick), through more efficient designs, and by increasing the efficiency of the fermentation process (Prasad et al. 1974, p. 1355-1356, discuss methods of reducing construction costs). Ideally, it should be possible to simulate the likely balance between the costs and benefits that would be associated with any change in plant design or operating procedures. For instance, the net benefit of increasing the temperature of the fermentation process might be compared with the net benefits of a longer material retention time, insulation of the plant from the surrounding air, agitation of the material being digested, preprocessing of the inputs, or the addition of supplements such as urea and urine, etc. Similarly, the unit cost of gas production might be tabulated for a variety of different plant scales and designs. However, neither the biochemistry of the continuous process nor the costs associated with such changes seem to have been established with sufficient accuracy to enable such precise comparisons to be made.

The Need for More Appropriate Data

The data that currently exist on the viability of biogas plants are not only very unreliable (with considerable variation in the values assumed by different studies for similar parameters), but are obtained from a narrow range of possible plant designs and a narrow range of socioeconomic and agricultural zones. The practicality of biogas plants will vary considerably among different locations according to the balance among the various costs and benefits operating in each location. It is therefore essential that evaluations are carried out in a much wider range of circumstances and using a number of improved designs. In Korea, all 23000 plants said to be operating are all the same size (about 35 ft^3/day, Mardon 1976, p. 16); in India most plants are between 150 and 200 ft^3/day (even though plants up to 5000 ft^3/day are offered by the KVIC (Moulik and Srivastava 1975); and in the Philippines the most successful plant is said to be the 1000 ft^3/day batch process plant (Philippines de la Salla University; Mardon 1976, p. 18-19).

Such studies might adopt a method of analysis similar to the one outlined here, and particular attention should be given to questions of the social acceptability of the production techniques. The researchers should involve the people who will be affected by biogas investment as much as possible both in the research design and in carrying out the research itself. Attention should also be given to the acceptability of the gas for cooking and to the acceptability of handling dung and other wastes.

Research on Community-Scale Plants

The income distribution in most developing countries, combined with the current costs of biogas plants, has led to the fear that family-sized biogas plants will worsen the income distribution of rural areas and will in any case only satisfy a very small proportion of total rural energy needs. If income distribution and the need to satisfy the fuel requirements of a larger proportion of the community are important it is argued that community-based plants are the only option.[18]

The number of community-based plants (as opposed to large plants run by institutions or agricultural industries) is thought to be very small, and there is therefore a particular need to research the issues surrounding such plants in varying social situations.

[18]Farvar and Bajrachanya (1975) stress that for reasons of income distribution and promotion of community participation, community plants are really the only types of biogas plants that would be promoted. This view is also put forward by Government of India (ICAR) 1976, Moulik and Srivastava 1975, and Prasad et al. 1974.

Biogas Systems in Asia: A Survey

S. K. Subramanian

India

Nearly 70% of India's biogas plants, which now total more than 36000, were built during the fuel and fertilizer crises of 1975-76. However, the Indian Council of Agricultural Research (ICAR) had begun anaerobic cow dung ('gobar' in Hindi) fermentation as early as 1938-39 (Patel 1975; Sathianathan 1975). Significant biogas plant use began in 1951, when the gasholder and digester were combined into one semicontinuous unit. The design most commonly used was introduced in 1954 by the Khadi Village Industries Commission (KVIC) and incorporated a device to stir the slurry and break up the scum.

The KVIC design offers capacities of between 1.5 and 85 m^3 of gas output per day at an estimated cost of between U.S.$260 and 4400 (KVIC 1975). However, most plants are for domestic purposes and have 3-7 m^3 capacities. To reduce costs, these designs are unheated and unstirred, and thus require a depth of nearly 4.5 m and a retention period of about 55 days. In the larger (85-m^3) units 40% of the gas generated is consumed for heating (Gobar Gas Research Station, Ajitmal).

KVIC is experimenting with larger 140 + m^3 units to gain experience for still larger ones. Projects by other groups include the Rural Electrification Corporation's 125-m^3 unit, Karimagar District, Andhra Pradesh (probably incorporating a 3.5-kW gas pump to recirculate contents three times) and a 420-m^3 digester for the Delhi Dairy Corporation.

The study reported in this chapter was undertaken by the Management Development Institute (MDI) at the request of IDRC, and was recently published as a monograph by MDI. Their permission to include this edited version in our book is gratefully acknowledged.

A grant from the Indian Council of Social Science Research, New Delhi, made possible visits to over 70 Indian biogas establishments and discussions with officials, extension agencies, plant owners, banks, etc., while IDRC financed visits to biogas plants and meetings with officials in Thailand, Indonesia, the Philippines, South Korea, and Japan.

Other Designs and Approaches

Early KVIC attempts to use split bamboo as a digester construction material in West Bengal failed because they were attacked by rats. Some 10 years ago, narrow (1-2.5 cm) earthen rings were successfully used to build a digester in Kalimpong, and three prefabricated rings, 1 m × 2 m in diameter, are currently used at a housing cooperative in Sangli, Maharashtra, for a 3-m^3 digester, and four for a 4.5-m^3 digester.

A number of new designs have been published by the Gobar Gas Research Station, Ajitmal (Rambux Singh 1973). One uses agricultural waste insulation and an external water jacket heated by a solar heater that delivers 1.5 litres of water per minute at 60 °C. It is claimed that by using this process slurry retention time has been reduced from 50 to 55 days to between 15 and 18 days, and gas production has risen by 300% even in the winter.

The steel gasholders of the current Indian design were found to work best despite their high cost (35% of the total capital costs) and maintenance problems (corrosion). Painting once a year (or monthly with engine oil) is recommended to prevent corrosion; at Urlikanchan, Maharashtra, 3 litres of engine oil are added to the top of the digester each month for this purpose.

Alternative ferrocement gasholders were found by the Indian Institute of Technology, Madras, to be too heavy (producing gas pressure of 20 cm of water) and to have poor strength and flexibility; furthermore, they could not be easily leak-proofed if a hole was bored through them. An experimental gasholder of woven bamboo, aluminum foil, and a polythene covering was tried, but it collapsed in a dust storm. Other local construction materials are being tested at the Indian Institute of Science at Bangalore. Early experiments in India with negative pressures (-1cm water) were not successful.

Operation and Maintenance

In the Indian design the dung:water ratio is 1:1. The design uses a relatively small amount of water and there is good mixing due to the high height:width ratio (6:1) and because the gas bubbles that rise from the bottom of the plant prevent settling out of the sludge (Mardon 1976). Of the plants visited during this study, 89% were in operation; some for over 10 years. This improved situation was due to the extension of service facilities. Recommended maintenance procedures include daily feeding and agitation, and annual painting of the gasholder. Plant failures were due to masonry construction defects, failure to paint the gas holders, improper feeding, lethargy of plant owners, and changes of ownership.

Winter Operation

The low winter temperatures (0 ^{0}C) of Northern India can cause gas production to drop by 20-30%. Farmers cover gasholders with plastic sheets after sunset during the cold months, and the addition of molasses, algae, urea, or urine is claimed to increase gas production. Authorities in Haryana State recommend building plants one size larger than needed to overcome the problem of low gas output.

Night Soil

The KVIC recommends connecting toilets to cow-dung digesters, as this conserves expenditures on septic tanks. The Gandhi Samarak Nidhi Institution also advocates night-soil-based plants and 55 such units operate in Maharashtra alone. Examples include a 14-m^3 plant using the night soil of 187 inmates of a leprosy home near Poona; two (10 and 4 m^3) units at the Parasakthi College for Women near Tenkasi, Tamil Nadu; a large pilot project of the National Environmental Engineering Institute based on 1000 inmates at the Nagpur Central Jail; and a 5.5-m^3 experimental plant in the Ratnagiri bus station, Maharashtra.

Other Wastes

Experiments on the anaerobic digestion of grass, water hyacinths, and rice straw have been conducted, the latter unsuccessfully due to choking. Distillery and strawboard mill wastes have been studied by the National Environmental Engineering Institute, Nagpur. In another case, using Hungarian technology, the National Sugar Institute, Kanpur, treated semiwet bagasse mixed with 3% cow dung and 5% urea, using city sewage as an initiator. This process has operated since 1963, and involves a battery of 12 batch digesters (6 m × 3-m diameter). After an initial 3-4 day aerobic treatment, the wastes are digested for about 40 days. A similar project using a wet process at the Aarey Milk Colony near Bombay produced gas at a rate much below the design capacity and proved to be uneconomical.

Gas

As most of the units are family-size, the gas is used for cooking: per capita consumption was observed to be about 0.2 m^3/day and much loss seemed due to inefficient burners. Specially designed burners can overcome the low pressure and flame propagation of CO_2-diluted methane, although coal-gas, LPG, and homemade tin-can burners are also used. At Parasakthi College, Tenkasi (Tamil Nadu), two plants

(17 and 28 m^3) feed an efficient boiler, generating enough steam to cook for 730 people.

Biogas use in diesel engines using a biogas to diesel fuel ratio of 85:15 has increased. The Institutes of Technology at Madras and Bombay, and the Indian Oil Company Research Centre at Faridabad are researching other engine applications.

The Tulsi Shyam Temple at Gujarat has (since 1966) used a 85-m^3 plant based on 300 cattle to run an engine that drives a water pump and a flour mill, in addition to generating 7.5 kVA of power for 4 hours at night.

Industrial uses of biogas are limited, but examples include small-scale KVIC soap and safety-match projects, and a water heater at a laundry near Bombay (42 m^3).

Slurry

There is a growing interest and emphasis on the manurial value of the digested sludge, and it has even been suggested that biogas plants would be more correctly named 'bio-fertilizer' plants. If the digester is close to the fields, the slurry is fed directly into the irrigation channel, but most often there must be several drying pits, which are used in turn. Many farmers add grass, straw, and bagasse to the slurry pit, thereby speeding up the composting process to about 3 months (instead of 9-12). In contrast to farmyard manure, slurry breeds no white ants and contains no weed seeds; a further advantage is that, unlike dung, the slurry is not stolen for fuel.

Vegetable farmers use digester slurry alone; others mix it with chemical fertilizers. Experiments at the Lalit Garden near Calcutta, West Bengal, comparing vegetable growth using compost, chemical fertilizers, and slurry found taste and size, especially of peas, best with the slurry. Weight of root vegetables increased by nearly 300% with night-soil slurry fertilizer compared with normal irrigation practices at the Central Jail Nagpur, Maharashtra. Similar success with Napier and Tara grass crops was reported by the V.S. St. John's Secondary School, Gannavaram, Andhra Pradesh, and with sugarcane at Digras, Maharashtra, and Katur, Andhra Pradesh. In both these cases the plants were built essentially for the value of the sludge.

Slurry is claimed to be ideal for nurseries and it has been used to correct the overuse of chemical fertilizer in rice fields (Dinikaki, West Bengal). Slurry has also been used successfully as a direct fish feed (after dilution) in West Bengal.

Integrated Systems

India has no system that attempts to integrate the slurry with the growing of algae, which in turn can be used to feed fish etc. Thick colloidal cow-dung slurry does not easily separate into sludge and a supernatant clear layer, possibly because of the restricted use of water for dilution in India. Even night-soil slurry has to be gravel-filtered at the Nagpur Central Jail where it is proposed to use the filtrate to feed algae and fish, and to irrigate crops. Using the slurry, Auroville Centre, Pondicherry, Southern India, plans to grow water hyacinth for banana plantation mulch.

Community Plants

Although there are a number of large biogas plants in India, none can be said to be a truly community plant. However, one Indian 'mini' community system operated between 1969 and 1970 in Khiroda Panchayat, near Bhusaval, Maharashtra. In this system several public toilets fed three digesters (5.5, 14, and 25 m^3) and the gas provided light for two city streets. Failure of the system was attributed to the transfer of the key operators and the electrification of the village.

KVIC and the Rural Electrification Corporation, together with the Council of Scientific and Industrial Research, are planning community plants at Digras, Maharashtra, and Karimnagar, Andhra Pradesh. The Digras plant will be fed by 20 animals and 10 community toilets; each family will be charged one rupee per month for the use of the toilets, generating an income of $11/month, and the gas will be sold to ten families. The slurry will be given back to those who supplied the dung in proportion to the number of cows owned.

The Karimnagar plant (125 m^3), based on 300 cattle, will provide gas to half the village (30 families), and run five 3.5-kW pumps, The system will be administered by the village and will employ two labourers.

Three semicommunity operations exist at the VSF Cooperatives, KCP Sugar Factory, at Vuyuru, Andhra Pradesh, the Madhavaram Dairy in Madras, and at the Kasturba Gram Krishi Kshetra, Indore. The first, a 35-m^3 plant based on 70 buffalo and calves (and ten baskets of sugar press mud daily), was built for $2200 (using a 25% government grant). This system supplies gas to 14 families for 3 hours daily at $2.25/month, one full-time labourer is employed and the system is considered uneconomic. The second (14 m^3) supplies seven houses (50 occupants) with 11 hours a day (nine in winter) for $1.75/month, again a full-time labourer is employed. The last, a 70-m^3 plant, supplies gas to 40 families 24 hours a day during February-July, and 14 hours a day for the remaining months. It is based on 200 cattle that produce 1600 kg of dung daily.

Minimum Number of Animals

The number of animals needed to support small domestic digesters is a key criterion in assessing the acceptability of biogas systems. It is usually suggested that five cattle are required for a 1.5-m^3 plant; this could put the technology out of reach of most Indians. However, this survey suggests that some 1.5-3 m^3 plants can be operated on two cattle or on an attached toilet and one animal because per capita demand with good burners can be as low as 0.08 m^3/day.

Dung output depends on the animals' feed and breeding, but the quantum of gas output is currently considered to be about 0.06 m^3/kg of dung. The KVIC now recommends 2-3 animals for a 2-m^3 unit, and 3-4 for a 3-m^3 plant.

Extension and Credit

The KVIC has played a key role in the extension of biogas. It has technical staff posted in all states to offer free expert advice. In response to an increasing workload the 'supervision charge scheme' was devised, and about 400 approved local artisans canvass potential customers, assist in construction, and help secure loans in return for a set fee from the KVIC (usually about $20 plus $2.50 per toilet). Today most State Boards also have their own technical staff, and since 1973 some State Agricultural Departments have been mobilized for the biogas program. In Haryana the entire District Government has been involved.

Until 1973 the KVIC gave grants of 50-70% to institutions (100% for "backward" areas); individuals got a $35-42 grant according to need, and an interest-free loan of up to $285 repayable over up to 10 years. This meant that $88-94 had to be provided by the plant owner who ended up paying between 24 and 52% of the total construction cost.

Commercial banks and State Agricultural Departments entered the field in 1973, and the subsidy was limited to 25% with KVIC giving the grant to institutions and the Ministry of Agriculture making grants to individuals. In 1976-77 the subsidy was reduced to 20%, and 5% decreases are planned for each of the next 2 years, after which it may be withdrawn. Banks provide the balance (up to 100% of the remaining cost) with a 4-year loan at 12-14% annual interest, on the basis of a mortgage, or personal and third-party guarantee. Community and "backward-area" installations will likely continue to receive liberal subsidies.

An interesting example of the provision of extension services in the cooperative sector is the Cooperative Sugar Factory at Sangli, Maharashtra. Biogas units have been built for sugarcane grower shareholders through a building cooperative using prefab structures. The mill guarantees bank loans and the advance is recovered from payments against the sugarcane crop. The scheme has existed for 2 years and will soon be extended to nonshareholders.

Recently the Government has launched the "All India Coordinated Project" on biogas for an integrated development of technology. The Government, the Reserve Bank, and other management institutions are concerned about meeting the 1978 target of 100000 plants.

The Republic of Korea

Nearly 27000 small digesters have been installed in Korea since 1969 through the efforts of the Office of Rural Development (ORD). However, the cold winters and lack of cattle make Korea's experience with biogas quite different from India's. ORD estimates that the country's severe winter results in national average fuel requirements of 43% for heating and 53% for cooking (ORD 1976). A rural household consumes 3.5 tonnes of farm products, 2.3 tonnes of firewood, 200 coal briquettes, and 20 litres of kerosene each year. Home heating is by the traditional 'ondol' under-the-floor system for which most of the rice straw, barley waste, and wood are burned: much deforestation and loss of compost material has resulted.

Research, development, and extension are handled by the Institute of Agricultural Engineering and Utilization, the Rural Guidance Bureau (under ORD) at Suweon, and the College of Agriculture. ORD also conducts experiments and is working on a large heated digester under the Korea-UK Farm Machinery Project.

Scale of Operation and Design

All the field units are of household size and consist of a rectangular underground concrete tank with an overflow, a feed pipe, and a mixing tank; digester capacity is 5.5 or 8.0 m^3. The 0.1-cm-thick PVC gasholder (later models have four compartments) rests inside the digester. The rectangular design is being changed to a circular one and a steel gasholder may be adopted because of the rise in price of PVC; the PVC holder costs \$55 (as opposed to \$65 for steel) and it deteriorates in sunlight. The price of the whole 5.5-m^3 unit is low at about \$140. Wooden gasholders with a plastic lining were abandoned because they leaked.

Operation and Maintenance

Cattle or pig dung (sometimes with night soil) is fed monthly or weekly at a 1:1 dilution. For a 20-day retention at 30 °C, the manual (Institute of Agricultural Engineering and Utilization, 1976) estimates gas production at 200-240% of digester volume. But actual gas production for the 5.5-m^3 model was 0.3 m^3 in January and 2 m^3 in September, or a 36%-of-given-capacity peak production.

Winter Operation

Most farmers do not operate the digesters between December and March, when temperatures are as low as -17 °C, and gas production is even inadequate for cooking. The gasholders are covered with straw during these winter months. Vinyl covers were tried but were ineffective and furthermore the sophistication of heating the digester was not justified for the small plants. Operation is more favourable in the warmer South.

Large-Scale Heated Digester

ORD has embarked on the development of village-scale digesters. A 40-family, 155-m^3 digester was operated at the Livestock Experiment Station, Suweon, under the Korea-UK Farm Machinery Project during 1976. The plant is based on 2.4 tonnes of dung from poultry and 170 cattle and has a retention time of about 40 days. The dung is mechanically mixed with water and urine in a separate unit (solid:liquid ratio of 1:2); 33-40% of the gas generated is used to heat the digester to maintain a temperature of 35 °C year round. In winter, hot supernatant liquor from the digester is used to melt the ice that forms in the mixing unit.

The primary digester is about 6 m in diameter, and is almost totally underground; the top surface is well insulated. A secondary digester (6-m diameter by 4 m) supports the gasholder, which has a capacity of 110 m^3. The gas pressure is 10-15 cm of water column. The gas is compressed and recirculated by a 'bubble gun' through the primary digester, and this breaks up any scum buildup. Experiments are continuing on the use of biogas in kerosene engine applications and home heating.

The total construction cost of the plant was $16000; $9600 for structures, $4000 for steel pipes and the gas holder, and $2400 for machinery and instruments. The ORD is planning eight more units that will be located in villages.

Other Developments and Inputs

The Institute of Agricultural Engineering and Utilization is experimenting with PVC and concrete fixed-dome digesters. The College of Agriculture at Suweon is working on a two-stage digester of reinforced plastic insulated with paddy husk. The plant is designed primarily for pig manure at 1:5 dilution and a 30-day retention; interestingly the gas passes through an algal culture to use up the CO_2 (Lee and Kim 1975). A primary school in Kyong Ju-Shi is operating a plant with night soil and the army has shown interest in this, although there appears to be psychological inhibitions against its use. The digestion of vegetable wastes has received scant attention in Korea (Lee and Kim 1975).

Gas

No farmer is totally dependent on biogas: it supplies only 3-6% of home heating, and less than half the cooking needs (43-45% as a family of five needs 0.7 m³ of gas for 3 hours daily). Biogas saves about 226 hours of housework per family per year (ORD 1976). Thus each house has a cooking fire (Mardon 1976) in addition to an unmodified LPG burner for biogas. However, heating, cooking, and power will be provided by the larger village-sized plants.

Slurry

Slurry in boxes is carried by hand to the nearby fields, mixed with compost, and applied; the sludge acts as soil conditioner. However, at the Livestock Experiment Station pilot digester, the slurry is diluted after settling and pumped directly to the fields. No special emphasis is placed on the manurial value of the slurry, and it is not taken into account in the evaluation of the pilot plant. Future plans involve the building of oxidation and algal fish ponds.

Minimum Number of Animals

A 4.5-m³ digester is said to need the total waste output from 8 cows, 23 pigs, or 630 fowl according to the design manual of Institute of Agricultural Engineering and Utilization. But often only two cows and a few pigs supply the household plants; therefore, gas output is often handicapped.

Extension and Credit

In each subregion ("gun") the Rural Guidance Office of the ORD (Ministry of Agriculture and Fisheries) provides technical extension and financial loan assistance to farmers, but there is no regular loan system and the 33-50% Government grant system has been discontinued. Most of the biogas construction in Korea is undertaken by the farmers themselves.

Rapid urbanization and the shortage of animal waste slowed the construction of family units in rural areas during 1975 and barely 4000 were built. In 1976 the ORD abandoned family-unit installation to concentrate on the development of village-size units, gas storage and purification, power generation, etc.

The Philippines

Fuel is not a major problem in the Philippines as firewood is plentiful. Consequently, interest in biogas stems from its pollution control and public health applications. Pigs (and some buffalo) provide most of the animal wastes, but despite some possible psychological inhibitions the National Housing Authority (NHA) is also promoting night-soil digestion, and one digester is already operating. Techniques to avoid night soil overdilution and to screen out harmful detergents have been developed. Further units are being considered for a Manila hospital and for the proposed Palawan Island Resort.

The major research activity is centred at the National Institute of Science and Technology (NIST), at the University of the Philippines at Los Baños, and Maya Farms. The greatest potential seems to be in the digestion of agricultural wastes as their volume is estimated to be 1000 times greater than livestock wastes.

Field Experience

Nearly 100 Taiwanese-type units have been built under NIST guidance. They cost over $690, and need the manure from 5 to 10 pigs diluted at 1:3. These units produce enough gas to supply the cooking needs of five people. Some other plants have been built by individuals on their own initiative. A NIST-prepared culture of 10 methanogenic isolates is recommended as a starter. In the Philippines, the digested slurry is not used to any extent as a fertilizer.

The Chan-type digester from the South Pacific has been adopted lately as it allows more room for maintenance. Prefab digesters and galvanized iron gasholders have also been experimented with by NIST. New digesters both for integrated systems and for agricultural wastes (straw, banana leaves, water hyacinth, etc.) are being worked on. Findings include the discovery that banana leaves or straw that have previously been used for mushroom growth are more readily digested.

Integrated Systems

The University of the Philippines at Los Baños has an integrated biogas system that uses slurry to grow *Chlorella*, fish, and rice. Waste from 10 pigs is diluted in the proportion of 1:4 and fed to two digesters in series (21-day retention). The gas would be sufficient to meet the cooking needs of five people. The slurry is settled out in two settling tanks, positioned in series, diluted, and channeled to the algal fish pond and the rice fields. A windmill stirs the algal culture and transfers liquids.

Maya Farms

Maya Farms (40 miles south of Manila) is the largest Asian biogas establishment, with 48 large (2.5×3×3 m) batch plants based on 7500 pigs (soon to be increased to 15000). Every other day, a digester is fed with 5 tonnes of a dung (dilution 1:1) and some predigested slurry as a starter. The contents are then stirred mechanically every day for 2 minutes. The gas (rich in CO_2) produced in the first 3 days is purged; after this, gas production is ideal for 23 days, at 60-80 cm water column. On an experimental basis, paddy straw is mixed with the dung.

In addition, five continuous digesters of Indian and Taiwanese designs are operated, producing gas at pressures as high as 45 cm water column; straw digestion experiments with these plants failed due to clogging.

Daily gas production for the whole farm is about 560 m^3, although the technical capacity is 840 m^3, but this would require more animals. The gas storage capacity is about 140 m^3.

The gas is used in a canteen (daily per capita consumption is 0.1 m^3), a meat processing plant, and a soup cannery. The gas also powers a 625 litre/minute water pump with an old 35-kW car engine, which has a consumption of 0.6 m^3/kW/h of gas. Similarly, a 110-kW car engine has operated a 60 kVA generator (1800 rpm) 4-5 hours a day for 6 months to run four freezers. Generally, biogas-run gasoline engines were found to give higher-than-rated rpm's, though with diesels the rpm's are lower. Refrigerators with a gas consumption of 0.08 m^3/hour, water heaters, lamps, burners, etc. are also on display. The burners are LPG models with the holes enlarged to 0.3-cm diameter. Steam generation has been abandoned because of low efficiencies.

The sludge from the few continuous digesters is fed directly into irrigation water, but the batch-digested sludge is settled out for 10 days, dried (with heat from biogas in rainy weather), and used as soil conditioner for submarginal soils. The liquid from the settling lagoon along with wash water is aerated for 7 days at 75 psig of pressure. A windmill will soon be used for the air compression in the aeration process. After BOD, COD, salt, and water plant growth analyses, the solution is used to fertilize the rice crop and grow *Chlorella*, which feed both fish and animals; feed waste is also added to the fish pond. The treated solution has been an effective fertilizer despite a low (1-2%) nitrogen content (except for an excess introduction of copper) and in fact, overfertilization has often resulted. Reduction of the treatment cycle of the digested slurry/sludge is being attempted.

A network of satellite farms has been established to minimize the risk of infection among the animals. The satellites consist of units of 25 young pigs that are supplied together with biogas technology to smaller farms in the area; 25 such units at eight local farms exist at present.

Extension and Credit

Responsibility for extension work is divided among the National Housing Authority, the Engineering Battalion of the Military, the Community Development Department, and others, although the NHA has a coordinating role in new settlements. The Development Bank of the Philippines recently began to give loans to pig farmers for biogas plants at 6% interest; approximately $412 (3000 pesos) is given for single and $550 (4000 pesos) for twin digesters.

Thailand

The Division of Agricultural Economics, Ministry of Agriculture, built a demonstration plant as early as 1965 (Deemark 1975), but subsequent development has been hindered by a shortage of livestock wastes. Currently the Department of Animal Husbandry at Kasetsart University, the Department of Health, and the Applied Scientific Research Corporation are all developing various biogas systems. The major constraints to the development of biogas appear to be the convenience of wood and charcoal as a fuel and the lack of dung.

Field Experience

There are now nearly 225 family-sized (2.8-m^3) units based on cattle or pig wastes in Thailand. The loading is 20-40 kg/day at 1:1-1:1.5 dilution; feeding varies from daily to monthly. The gas produced meets the cooking needs of between 5 and 7 people using homemade or LPG burners. Some farmers use the digested slurry on their gardens, but most use chemical fertilizer (ESCAP 1975).

The sanitation centre in Sara-Buri near Bangkok, in collaboration with the Faculty of Public Health, Mahidol University, is working on night-soil biogas research. Kasetsart University has also built four family-sized units on the campus for research and demonstration purposes. No community-size plants are contemplated.

Galvanized Iron Gasholders

Early galvanized iron (G.I.) gasholders built by the Division of Agricultural Economics, Ministry of Agriculture have had to be replaced by the more expensive steel models. But during the survey three satisfactory G.I. units were found in Lopbury Province, and in Ban Mee District galvanized iron coated with asphalt was used to build a 1.5-m (diam.) gasholder that has performed satisfactorily since 1973, for only 600 bhatt ($30), or half the normal cost. Leaks in another G.I. drum built for 400 bhatt ($20) were repaired with white lead and boat caulking cement, and one such plant has operated for 12 years. Interestingly, in this latter case, shade was claimed to improve digester performance.

Industrial Wastes

The Applied Scientific Research Corporation of Thailand is developing an anaerobic digestion process to treat about 1500 m^3 per day of high-BOD distillery waste (potential daily production should be about 47500 m^3 of biogas). This is primarily to treat waste and the methane will be flared off because impurities (especially H_2S) rule out distillery use, and scrubbing is considered too expensive.

Extension and Credit

Sanitation is the most important consideration in biogas installation in Thailand and consequently the Health Department has responsibility for its promotion, primarily to control disease carriers like fruit and house flies (Mardon 1976). Coordination is carried out by the Rural Development Department.

A sanitation officer from one of nine centres travels to the villages. Farmers who are interested in a plant make an initial payment, after which, delivery is arranged. Most plants have been built in Saraburi province with some units being installed in the homes of headmen for publicity. The iron molds used for gasholder construction unfortunately limit digester size.

A government subsidy scheme has now been withdrawn, but the Agricultural Banks are contemplating loans for biogas as part of a fertilizer scheme.

Indonesia

Only twelve units are said to be in operation in Indonesia as firewood is plentiful in most areas and animal wastes are not. Muslim opposition to pig dung use may also be a limiting factor as it forced the Indonesian Board of Voluntary Services (BUTSI) to move a plant built at Bakum.

Demonstration Units

Oil-drum demonstration units (digesters made of one or more empty oil drums) include a train of six double-drum poultry manure units at Denpasar. Two units are used at Petung — one heated with compost around the drum and another based on night soil. Various dilutions are being tried and the use of solar heating is being investigated by a civil engineer at Yogyakarta. A 7.8-m^3 rectangular unit produces cooking fuel at Atuag. In 1976 Community Aid Abroad (Australia) assisted a Bogar school to set up a $600 (250 000 rupiyah) 3-$m^3$ Chan digester based on 25% of the dung of 20 cattle — lack of water forced operation at a dilution of 1:2. Predigested slurry is used as a starter, and feeding is once every 2 days. The school has also experimented with a three-drum unit. There is great reluctance to use the gas, however.

The Development Technology Centre (DTC) at the Bandung Institute of Technology recently set up a 1 × 2 × 0.95 m rectangular unit with an overhead gasholder and a water-jacketed top at the Buruadjak dairy farm, Lebang. Feeding is weekly (dilution 1:2), and a glass window allows observation. Several triple oil-drum digesters are also operated by the DTC at Lembang. Removable connecting joints allow better maintenance.

Extension

The village technology unit of BUTSI hopes to promote biogas by demonstrating its feasibility to village chiefs.

The Bogor Biological Institute will soon launch a biogas program based on agricultural wastes — 31 million tonnes of corn and paddy stalks (four times the animal wastes) are available yearly (Sudirjo and Kismomihardjo 1975).

Night-soil use is likely to be accepted for digestion though the use of slurry as manure has yet to be, especially in Bali, where a witch doctor attributed sickness to its use.

Japan

Small digesters are said to have operated in the Tohaku region for many years. Recently several institutions, including the National Institute of Animal Industry at Chiba, the Public Works Research Institutes, the Fermentation Research Institute at Anage, M/S Hitachi Plant Construction, the Ministry of Agriculture, and the Agency for Industrial Science and Technology (MITI) have worked on anaerobic digestion of rural, urban, and industrial wastes for pollution control. Japan is the only country in the region to have adopted high-temperature digestion (in the thermophilic range) of some wastes.

Livestock Wastes

The growing Japanese pollution problem has resulted in a spate of antipollution laws and methods of meeting them. Since 1973 a multiinstitutional, nation-wide effort has attempted to reduce the pollution problems of animal wastes. The energy crisis added further impetus to this effort. Digester experiments include a 200-litre unit (60 cm in diameter, made of fibre-reinforced plastic insulated with 5 cm glass fibre) based on the wastes of one pig diluted 1:3 with a 16-day retention period. A 160-W submerged pump agitates the contents, and a temperature of 35 °C is maintained. Gas production is about 20 litres of 62% methane per day. Dry matter content of the slurry is 5.2%; organic matter content is 3.6% wet and 71.83% dry basis; nitrogen content is 0.32% wet and 6.19% dry basis (Yagi 1975).

A similar larger unit (5 m^3; 1.5 kW pump) based on 25 pigs at the Kagawa Prefecture is being researched by the Ministry of Agriculture. Digested slurry from this unit is fed directly to the fields (Yagi 1975).

A large Kochi Prefecture digester, which is insulated with vinyl, maintains a year-round temperature of 30 °C. Poultry-dropping digestion experiments have resulted in toxic ammonium ion accumulations (over 300 ppm). No night-soil digestion has been tried.

The Nippon Veterinary and Zootechnical College, Department of Animal Hygiene, found mesophilic digestion more efficient than thermophilic (Kamata and Uchida 1972), and the Public Works Research Institute found that sewage sludge contains heavy-metal contamination that is likely to render it useless as a fertilizer for edible plants.

Industrial Wastes

The Fermentation Research Institute, Inage, has been promoting thermophilic anaerobic digestion of industrial wastes (i.e. distillery, butanol, yeast, antibiotic, and paper mill waste) since the Second World War. BOD removal is 70-90%, and the sludge is used for fertilizer.

Both thermophilic and mesophilic digestion produce the same amount of biogas per unit of volatile solid, but the former allows reduction of the retention period to 5-7 days and loading rates 2.5 times greater (mesophilic loading rate is 2-3 g/litre/day versus 5-6 for thermophilic digestion). This allows a considerable reduction in digester size (Sonoda et al. 1965).

Twelve distilleries have anaerobic digesters; five working under government supervision treated 200 million litres of wastes in 1966 and recovered 170000 m^3 of gas for fuel (Fermentation Research Institute 1974).

Recently, digestion of distillery waste has been discontinued because it does not remove the brown colour pollutant. The 1000-2000-m^3 units at the Chiba Distillery, for example, have been converted into aerators, with the concentrated sludge being discharged into the sea.

Urban Wastes

These wastes are currently either incinerated or used as land fill. But, as part of the MITI "Sunshine Project" Hitachi Plant Construction has been investigating their optimum anaerobic fermentation (Takatani et al. 1975) and has concentrated, since April 1975, on thermophilic digestion only. The experimental units are small (1 litre, 100 litres, and 1 m^3), but a 1200-m^3 plant is being planned. The dilution is 1:2 and the C/N ratio is kept at 1/20 with ammonium carbonate. The retention periods are 25 and 7 days, respectively, for mesophilic and thermophilic digester. The latter has a loading rate approximately 2.4 times greater, and in winter one-third of the gas is used for heating.

Other Countries

Bangladesh

As alternative fuel sources in Bangladesh are limited, the Council of Scientific and Industrial Research and the Bangladesh Academy of Rural Development have built some demonstration plants. Polythene-bag designs are, in an attempt to reduce costs, being tried at the Bangladesh University of Engineering and Technology at Dacca (Islam 1976). Long-range plans include combined water hyacinth - cow dung digestion and village-size plants are also being considered as they may be suitable for the particular social structure (Eusuf 1975).

China

Biogas is extensively used for cooking, lighting, fertilizer, and for small internal combustion engines (Fang Chen 1976). As of September 1975, over 200000 family-size (10-m^3 capacity, generating about 5 m^3 of biogas per day) digesters were operating in the province of Szechuan (ESCAP 1975). They are built essentially underground, of brick, cement, and pebbles, with no moving parts; the gas pressure is kept constant automatically by changing water levels. A key factor is the size of the door connecting the fermentation tank to the outlet chamber. The summer temperature is about 23 °C, winter temperature about 10 °C.

Maintenance is carried out once a year (People's Publishing House 1974). The feed is a mixture of urine (30%), night soil (10%), and water (50%); vegetable matter is decomposed for 10 days before inclusion. Lime solution or grass ashes are added to maintain a pH of 7-8. The burners are made of soil and carbon ash, with a biogas:air ratio of 1:10. Free-standing biogas lamps are used (Production Team of T'ang Ngan 1973).

The Research Office for Parasitic Disease Prevention and the Revolution Committee of the Mien Chu District Communicable Diseases Prevention Office have found that the best pathogen control method is to remove digested sludge from the middle of the digester to allow worms and eggs to settle in the digester. Physical and chemical destruction of the worms and parasitic eggs is then carried out after 6 months retention when the digester is fully emptied (Research Office for Parasitic Disease Prevention, Province of Szechuan 1973).

Nepal

Since the first biogas plant in 1970, the Development and Consultancy Services of the Butwal Technical Institute and the Energy Research and Development Group under Tribuvan University have contributed to the construction of 100 2.8-m^3 digesters in 1975 alone (cost: $400 each). Major problems include transportation and steel costs, water access, and the provision of loan credit when fixed assets are limited.

Pakistan

The Government has built nearly 100 units, some capable of producing 11 m^3 of biogas per day. Gasholders are free if the farmers build their own digesters (ESCAP 1975). Demonstration units at military dairy farms, universities, and integrated rural development centres are built at Government expense.

Problems include low winter temperatures, waterlogged hilly areas, and high steel costs. The Appropriate Technology Development Organization designs and builds 10-m^3 fixed-dome digesters based on Chinese technology (cost per unit about $590, or 5600 rupees, ESCAP 1976).

Sri Lanka

A 2.8-m^3 demonstration plant was built by the Industrial Development Board, Ministry of Industries during 1973-74. Research and development were carried out concurrently by the Peradeniya and Katubedda campuses on vegetable material digestion (especially of salvina and water hyacinth). The Government plans to introduce subsidies to encourage the use of nonconventional fuels, and the Asian Rural Energy Research Project Experimental Station to be located in a rural village near Hambantota, which was UNDP-assisted, will also have biogas generators.

A cheaper and more compact generator with no moving parts is being developed by IDB and is named the "Lakgen." The main components will be two underground brick static tanks, with one at a higher elevation than the other, so that gas pressure is maintained by the slurry. In addition to low initial cost and ease of operation, the gas will be available at a higher pressure than in the current units (Industrial Development Board, Sri Lanka 1976).

Taiwan

In 1973 there were said to be nearly 7500 family-size biogas units on the island, most based on 12 hogs. The design includes a unique manual mixing device made out of PVC pipe tied to a piece of rope. The digested sludge is used as fertilizer, with a small part being used for *Chlorella* cultivation (Chung Po 1973).

Integrated systems, combining a bag digester with algal and fish cultivation, are said to originate from Taiwan (ESCAP

1975). The digester bag is a light, mass-produced bag of 0.55-mm hypalon laminated with neoprene and reinforced with nylon, with a PVC inlet and outlet. Small circular (5-30 m^3) and large rectangular bags (50-100 m^3) are available from Fortune Industrial Corporation, Taipei.

Interest of the International Agencies in Asia

Biogas systems are now receiving attention from several international agencies following the crisis in the supply of energy and fertilizer. After its 1974 Colombo Declaration, the Economic and Social Commission for Asia and the Pacific (ESCAP) held biogas workshops (in New Delhi, August 1975, on Technology and Economics, and in Manila, October 1975, on Fermentation Technology), and began publishing newsletters. The Energy Division of ESCAP will survey the potential of energy resources in the region (including biogas) in 1978.

UNIDO has asked UNDP to finance a global project on biogas plants during 1978-1982, but the future of the project is still uncertain. UNICEF and WHO have also expressed interest.

Under UNEP's Rural Energy Project, pilot digesters are being built in Senegal and Sri Lanka, in cooperation with the Brace Research Institute, McGill University, Canada, and Oklahoma State University, USA. A 84-m^3 unit in Sri Lanka is to run a 6-kW generator for lighting and water supply. UNEP is presently considering a request from Kenya for assistance in harnessing solar, wind, and biogas energy in a rural area. Through its International Referral System (IRS) UNEP is also facilitating information exchange on the subject. The World Bank has expressed interest, but believes that a thorough analysis of the technical, economic, and financial feasibilities must yet be made.

Some Generalizations

Within the limited efforts made so far in Asia, there is already a considerable diversity of successful systems suitable to conditions in the country of application. This should be recognized in planning future programs, as emphasis on uniform design may seriously reduce the potential usefulness of the technology, which may well depend on being adapted to the detailed features of the location in which they are to be used.

The number of viable systems may be even larger than those established to date. Thus experimentation under local conditions must be fostered, although it may well be possible in the context of highly standardized (perhaps mass-produced) parts, components, and materials.

Technological Aspects of the Region's Experience

The socioeconomic factors governing biogas systems are very much interrelated with the technical factors. Higher digestion efficiencies reduce plant cost; whereas, cheaper and more easily accessible inputs and the efficient use of outputs can bring the system within reach of a larger section of the rural community. Some of the technological findings of the survey are presented here.

Biochemical and Other Operational Aspects

Poor Digestibility of the Dung

One kilogram of volatile solids could produce 0.75-1.0 m^3 of biogas at normal temperature and pressure (NTP), depending on the quantity of carbohydrates, fats, and proteins in the feed. Although dung contains 75% volatile solids (dry basis), the digestion efficiency is only about 20% (cow dung produces 0.09-0.2 m^3 biogas/kg, volatile matter; sewage sludge 0.4-0.6 m^3, Mohan Rao 1974). Most of the lignin-bound cellulose is not digested.

The efficient digestion of cellulose depends on its rate of hydrolysis into sugars. This can be achieved by heating under pressure, or treatment with acids and bases. However, these are not applicable at small scales, and the only possibility at this scale is a relatively low-cost cellulase enzyme. But there has been little research on hydrolysis even though it is largely responsible for the slowness of the digestion process (see Chapter 1). Hitachi and the Nomura Research Institute, Kangawa, Japan, are studying some problems of hydrolysis in paper-waste digestion, but it should be noted that increased digestion reduces the humus content of digested slurry though the nitrogen content remains largely the same.

Use of Other Inputs

The field experience of the Asian region (except Japan) is essentially confined to the treatment of livestock wastes and night soil; however, the digestion of fresh and dry plant residues, algae, and various marine, agricultural, and biological wastes is being investigated in pilot plants. For example the National Sugar Institute, Kanpur, India, has been operating a complex agricultural waste system. Water hyacinth and algal research has been initiated in Bangladesh, India, and the Philippines and nearly 1.9 ml of biogas/g of water hyacinth have been obtained; cadmium and nickel contamination actually increases production (NASA 1974). Evidence exists that pretreatment of agricultural wastes (such as chopping, soaking, decaying, or mushroom cultivation) assists their digestion (NIST, the Philippines). Where livestock wastes are scarce, the digestion of these other organic wastes becomes very relevant, but little data currently exists. More research in this field is necessary.

Frequency of Feeding

On the basis of the Indian and Chinese experience, the secret of successful biogas-plant operation lies in the daily feeding cycle. Ideally, but impracticably, the feeding should be continuous. The National Environment Engineering Institute (NEERI), India, resorts to feeding three times a day, and some farmers in India practiced twice daily feeding in winter. Outside these two countries, however, daily feeding has not been generally adopted.

Organic Loading

The production of gas is also dependent on the weight of volatile solids added per digester volume per day. The size of the digester consequently depends on the loading, which in turn depends on dilution, retention time, and temperature of digestion. Maximum loading is 2-3 $kg/m^3/day$ in mesophilic digestion, and 5-6 kg in thermophilic. Loading rates in India (mesophilic) are around 1.6-2 $kg/m^3/day$. Loadings of 3.17 and 3.2 $kg/m^3/day$ have been achieved (Yagi 1975; NEERI, India).

A further 2-300% increase is possible if the sludge concentration is increased to over 10% (Fermentation Research Institute; Sonoda et al. 1965). Loadings of 1.8-7.6 kg for mesophilic and 1.8-18.8 kg for thermophilic digestion are reported (Kamata and Uchida 1972). Hitachi of Japan (urban wastes) reports rates of 0.77-4.7 kg and 1.73-12.6 kg for mesophilic and thermophilic digestion, respectively (Takatani et al. 1975).

Dilution and Retention Time

These two factors are interdependent and the experience of the region varies considerably. India uses a 1:1 dilution and a 50-day retention period though a 30-day retention is thought possible even with existing designs. At Maya Farms (Philippines) a 1:1 dilution is standard, with a 45-day retention soon to be reduced to 23 days; spend slurry starter is used. The University of the Philippines uses a 1:4 dilution and a 21-day retention; NIST dilutions are 1:2-1:3. The Agricultural College at Suweon, Korea, normally practices 30-day digestion at 1:5 dilution; this is reduced in their heated digester, which receives feed at 1:2 dilution for a 20-day retention. The National Institute of Animal Husbandry (Japan) dilutes feed to 1:3 for a retention period of 16 days. Hitachi dilutes urban wastes with sewage sludge (1:2); mesophilic digestion takes 25 days, thermophilic takes 7 days.

Higher loading rates would reduce digester volume, cut down on the heat load and water requirements, and minimize sludge disposal problems.

Mesophilic and Thermophilic Operations

In the region, high-rate thermophilic digestion is practiced only in Japan for industrial wastes that are discharged at high temperatures. At Hitachi, the gas generated under both mesophilic and thermophilic conditions was similar at between 320 and 340 ml/g of volatile solid, but the organic loading rate was 3.8-3.9 g/ml/day for mesophilic digestion and 9 g/ml/day for thermophilic — a 2.3-fold increase resulting in a considerably reduced retention time (Takatami et al. 1975). Since April 1975, Hitachi has used the thermophilic process only.

In a comparison of mesophilic and thermophilic digestion, the Department of Animal Hygiene, Nippon Veterinary and Zootechnical College, found the thermophilic process superior during the first half-period, and the mesophilic more efficient over the second. They concluded that the cheaper mesophilic process was more appropriate, taken overall, for the treatment of pig feces (Kamata and Uchida 1972).

Carbon/Nitrogen Ratio

The optimum C/N ratio is usually given as 1:30, but the N and C content of the feed varies with the age and growth of the feed plants, and the diet, age etc., of the animals. Moreover, what is measured chemically is not what is available to the bacteria. For instance, the digestion of some vegetable wastes failed due to a lack of nitrogen (NIST, the Philippines). Nitrogen-supplementing additives include ammonium carbonate (Hitachi, Japan where the C/N ratio is maintained at 1:20, Takatani 1975), 3% urea (National Sugar Institute, India), cattle urine, molasses, oil cakes, or algae. A 22% rise in gas production was said to result from a 1% addition of algae (National Dairy Research Institute, India). But some Indian scientists have doubts about the importance of nitrogen in anaerobic digestion because the slowness of the process means a lower nitrogen demand. Algae benefits may be partly due to easy digestibility, while the urea may only increase CO_2-production. Furthermore, the ammonium ion was reported to be toxic.

Removal of Toxic Materials

The harmful effect of certain toxic materials is known to cause the failure of digesters.

High concentrations of ammonia, lignin, certain essential oils (orange peel), H_2S, highly saturated alcohols, and some unsaturated alcohols have been found to be toxic (Fermentation Research Institute, Japan). Soluble sulfides formed by the reduction of sulfates affect the digestion of yeast wastes (removal results in a 40% rise in gas production, Sonoda and Seiko 1973). The toxic effect of unsettled undiluted slurry has also been noted and may be due to high BOD levels.

Kill Rates of Pathogens

A retention of more than 14 days above 35 °C seems to remove most pathogens. Reduction of the hardier parasite eggs appears possible through physical separation. Chinese studies report an 80-98% reduction in parasite egg concentration with an improved effluent storage chamber (McGarry 1976), and NEERI, India, claims 99% pathogen removal using oxidation-pond after-treatment of the sludge (a 30-day retention reduced hookworm egg concentration by 93% and roundworm egg incidence by 70%). Chinese studies show that for digestion periods of 10-90 days, the hardiest egg was *Ascarid* (roundworm). Egg viability ranged from 63 to 79% and remained at 47% even after 100 days. Paratyphoid B. bacilla survived for 44 days, and schistemes up to 37 days (Research Office for Parasitic Disease Prevention 1973).

Anaerobic digestion, therefore, compares well with any other feasible techniques for handling night soil and is clearly better than existing malpractices in excreta disposal. But, very low retention times can be a source of trouble (McGarry 1976).

Fertilizer Value of the Slurry and its Loss through Drying

Data on the fertilizer value of digested slurry are scarce and vague; anaerobic digestion does not create fertilizer, as some claim. Total waste solids are reduced, the nutrients concentrated, and the form of some of the nitrogen changed, but total nitrogen in the slurry is essentially conserved. The problem is to determine how much more nutritive the slurry is in comparison with the original material put into the digester, not just when it comes out of the digester, but also at the point of its end use. There is often a confusion of the meaning of slurry and sludge in certain publications: the first is simply the digester effluent; the second, the settled-out effluent with much of the liquid removed.

Fresh cattle dung contains 3.5% nitrogen (dry basis), 74% in organic form, 26% in the more assimilable ammoniacal form. Digested slurry contains 50% organic and 50% ammoniacal nitrogen (Hart 1963) — a 24% increase in the latter. Similarly, Acharya (1956) and Idnani and Varadarajan (1974) report that 15% of the dung nitrogen is converted into the ammoniacal form. The digested slurry contains 17.5% of its nitrogen as ammonia; in fresh dung ammonia makes up only 3.5-5.5% of its total nitrogen. NDRI, India, found 16-18% of the slurry ammoniacal. China reports approximately a 10% ammonia nitrogen concentration increase (People's Publishing House, Peking 1974).

But Leui (1975) claims that slurry contains 60-75% of physiologically active nitrogen as ammonia, 25% as amino acids, with the balance difficult to utilize. And United Aircraft Corporation (USA) found that digestion of cattle wastes increases crude protein by 100% and amino acid content of the digested product by 400% (Coe and Turk).

The ammoniacal transformation therefore appears to be dependent on food composition, and this could expain the variable data. For example, experiments on the digestion of cow dung, with the addition of various carbohydrates and 1% nitrogen (as nut cake), found that 43-63% of the total nitrogen was transformed into ammonia. With other organic materials (in combination with the dung), conversion into ammonia varied from 11.8% with bagasse, to 23.2% with legume leaves; it was 9.1% with dung alone (Idnani and Varadarajan 1974).

The drying of the sludge volatilizes over 97% of the ammoniacal nitrogen (ICAR, India) and results in a net loss of 18% of the total nitrogen. Dried sludge residue contains 1.78% nitrogen; if the ammoniacal nitrogen were conserved, the figure would be 2.16%. Nitrification studies on N-availability confirm this: with 30 mgN/100 g of soil, in 5 months the extent of nitrification was 21.3% for fresh slurry, 16.3% for compost, and 18.6% for sun-dried slurry (Idnami and Varadarajan 1974). Volatilization and loss are attributed to alkalinity and the pH during fermentation rises from 7.2 to 8.3, presumably because of ammonia accumulation.

To avoid the loss of the ammoniacal nitrogen, it would seem best to apply the slurry wet and plough it under; however, ICAR experiments on wheat, marua, etc., show that sun-dried slurry makes better manure (Idnani and Varadarajan 1974). Fertilizers were applied to crops on an equivalent nitrogen basis (125 kgN/ha for wheat); farmyard manure fertilized yields were taken as 100. The results for wet digested slurry were 103, for dry 113. When only 30 kgN/ha were applied, ammonium sulfate produced a yield of 137 (Berger 1976). Nitrification studies with 60 mgN/300 g of soil showed that after 3 months 7.4% of the nitrogen in the digested slurry, 4.7% of the N in the farm manure, and 87% of the ammonium sulfate nitrogen was nitrified.

Thus, the 'chemical fertilizer' ammonium sulfate is at least four times as available and effective as manure and slurry. With both of the latter, the nitrogen is only one-third available, with the rest becoming available during the second and third years after application (carryover effect). Chemical nitrogen is applied yearly and has little carryover (ESCAP 1975).

Damage to the soil through the repeated use of only chemical fertilizer is well known: during the survey, some fertilizer experts stated that organic and chemical fertilizers were complementary; and ICAR experiments show that incorporating chemical fertilizers into the slurry resulted in manures superior to either alone (Idnani and Varadarajan 1974).

Composting organic wastes with slurry results in a compost that is ready for application in 3 months, one-quarter of the usual composting time. This does not conserve nitrogen well, and ICAR experiments show that nitrogen efficiency is only 20-30% as opposed to 30-50% in regular composting (Idnani and Varadarajan 1974).

In experiments on facilitating the handling of the very dilute (90% water) slurry, Idnani and Chawla devised a filter bed of green leaves or straw from which semisolid residue could more easily be removed. The resulting mixtures were a better manure than compost.

The University of the Philippines at Los Baños claims that daily feeding of digested slurry to fields results in a plant nitrogen intake only slightly lower than with urea; Maya Farms even reported overfertilization with slurry alone. But the situation is complicated by carry-over effects and rates of application. Such remarkable results through the use of digested slurry/sludge are thought by some to be due to the action of humic acid on plant roots and to the presence of various micronutrients. Anaerobic digestion causes only a 20-30% loss of organic matter, thus plant residues (humus) are conserved.

For optimum utilization of digester nutrients, the feasibility of establishing an integrated farming system incorporating aquaculture as well as agriculture should be considered in view of the high yields and short life-cycle of biomass in water. Similarly, ways to use colloidal cow-dung slurry should be explored. Finally, the possible contamination with heavy (and other) metals and pathogens should not be overlooked in studies of the fertilizer value of sludge and slurry.

Design Aspects

Agitation

Agitation of the digester contents is often recommended to ensure intimate contact between the microorganisms and their food and to increase the rate of decomposition by releasing small trapped gas bubbles from the microbial cell matrix. It also helps to break up scum. Certain authorities claim that loading could be increased by four times in well-stirred, high-rate digesters (see Chapter 1). Most family digesters now in use make no provision for agitation. But there are exceptions such as the 140-m^3 digester being designed in India, the large-scale pilot unit in Korea, the batch digesters in the Philippines, and all units in Japan. Some Japanese models use a submerged pump for both agitation and mixing (Yagi 1975). Gas recirculation was reported to be both beneficial (India and Korea) and of little benefit (Japan).

Some experiments show that stirring may have only a temporary benefit on gas production. The Public Works Research Institute, Japan, found that continuous mixing produced only 5% more biogas than once-a-day agitation. But Hitachi, which adopted intermittent agitation in small-scale trials, uses continuous agitation in the larger models. According to them, the effect of agitation may not be apparent in small-scale digesters though it does break up the scum. The current practice in industrialized nations is toward continuous mixing (see Chapter 1).

Winter Operation and Heating the Digester

Reports on the operation of gas plants in winter are not consistent. Research institutions report a reduction of 60-72% in gas production during winter, but the State of Haryana, India, reports only a reduction of 25-33%. The problems of winter are overcome by setting up a plant one size larger than normal requirements, feeding larger inputs into the digester, covering the gasholder with plastic sheets, using hot water for feed preparation, and adding various materials like urea, urine, molasses, and oil cakes. Korean units experience a gas production drop of 85-90% during their severe winter. Part of the inconsistency may be explained by the greater efficiency of the larger scale plants under identical operating conditions (Hitachi Plant Engineering and Construction Co. Ltd. 1975).

Systematic investigations on digester heating have been carried out only in Korea and Japan. The heat could be supplied by burning biogas or by recovering waste heat from gas-operated engines. Percentages varying from 25 to 47.7% of gas produced have been needed to keep digesters in the mesophilic range; again, the different scales of operation may account for the divergence in amount of gas needed.

Experiments using solar heat are in progress in India at the Gobar Gas Research Institute, and in Indonesia. Another possibility, that of using heat liberated during aerobic composting to conserve heat in anaerobic digestion, was suggested as early as 1952 (Lessage and Abiet 1952).

Digester

The need for twin digesters is being questioned: both NEERI (India) and the University of the Philippines at Los Baños use twin digester systems and have concluded that a single digester could serve their needs just as well.

Defects that had caused failure of the digesters seen during the survey essentially concerned the masonry work and did not pertain to the design. The Indian design at times suffers from choking of the feed inlet. A problem with the Taiwanese design is the development of difficult-to-repair leaks in both the water seal and digester compartments.

Among the other designs, the Chinese model with its built-in gas dome and lack of moving parts has attracted attention in Pakistan, Korea, and Sri Lanka. In addition to delivering the gas at increased pressure, it is easy to construct in rural areas, and dispenses with expensive steel gasholders. But the annual maintenance and sludge removal could prove bothersome. The Indian design, in contrast, has been in continuous operation for over 10 years in some well-run plants. The other development is the bag digester: two brands, hypalon and butylon (Dunlop, New Zealand) are available. They have not proven as light nor, at between $255 and $1420 each, as inexpensive as was first expected.

Gasholder

In the Indian design, the steel gasholder makes up 30-40% of the total cost. With proper maintenance (i.e. annual painting) they have operated for over 10 years. Wooden gasholders in Korea failed and were replaced by PVC holders, which also developed cracks due to weathering. Later PVC designs have four independent chambers so that damage to any one does not affect the others. Ferrocement is relatively heavy and inferior to steel in strength and flexibility, but work on building low-cost gasholders from local materials is being continued in India. An external water jacket may be useful for a night-soil digester. Transparent gasholders have been suggested to increase solar radiation, but there is some doubt about how tolerant methanogenic bacteria are to light.

Utilization of Gas

Methane suffers a major storage problem as it does not liquefy under pressure at ambient temperatures (critical temperature and pressure: -82.5 ^{0}C, 46.0 bar). To store or transport the energy equivalent of 13 litres of gasoline as compressed gas at 2000 psi, a 1.6×0.27 m cylinder weighing 60 kg is required (Meynell 1976).

The low pressure of biogas and the low flame propagation speed of methane (66 cm/sec), which is further inhibited by CO_2, call for specially designed biogas appliances. Watson House Laboratory recommends that biogas burners have a total flame port cross-section area 300 times the injector cross-sectional area. Suitable flames can be obtained with orifices of 0.96 and 1.04 mm with a gas pressure of 2.5-20 cm of water. The heat output range varies from 130 to 430 kg-cal/cm^2 of port area/hour. Indian burners with a 60% efficiency use large 6-mm ports and the premix flame is short.

Biogas lights are generally inefficient, but the Chinese report the brightness of a standing lamp to be greater than that of a hanging one (Production Team of T'ang Ngan 1973).

Findings on the utilization of biogas engines are not consistent. A compression ratio of 13-15 is recommended for biogas use in engines (compared with a ratio of 6 or 7 for gasoline engines). Carbon dioxide increases anti-knock characteristics and does not have to be removed. India has had success toward adapting diesel engines to biogas using a biogas:diesel fuel ratio of 85:15. Engines running on biogas can go five times longer without an oil change.

Few examples exist of other systems of energy conversion using biogas, but Parasakthi College in Southern India effectively uses biogas to run a steam boiler for cooking purposes. Other recommended uses for large-scale operations involve the separation and use of CO_2 to make calcium carbonate, to promote algal growth, and to make dry ice for local health services, refrigeration, etc. (Pathak and Colah 1976; Prasad et al. 1974).

Future Research and Development

Technological problems to consider, in rough priority of order, include: the design of efficient burner and gas-use equipment, such as refrigerators and gas distribution systems; design of biogas-operated engines; and studies to determine the best use for slurry and sludge. Fermentation kinetics studies are necessary to find the optimum dilution, retention time, organic loading, etc. As well, improvement studies of dung digestion through enzyme action and other pretreatments should be undertaken.

Digester design itself should be scrutinized; different climatic conditions should be assessed and designed for; and the use of solar and wind energy should be investigated. Finally, the use of industrial wastes from agro-based industries, and the isolation and 'education' of bacteria for operating at low temperatures might be undertaken.

Social and Economic Issues

Decisions are being made all the time about biogas-related investments. Such decisions are made on both technical considerations and information about social and economic issues. As we have shown, technical data are often not available with sufficient accuracy and vary in different situations. This section describes those results of the survey of the Asian experience that relate to the different social and economic conditions, problems, and achievements of the region. Clearly, such a review cannot be a comprehensive treatment of all the issues involved, and this is not intended.

Owners

On the Indian experience, people who have so far been able to benefit from biogas plants have been in or above the middle-class levels. For instance, the survey carried out in Gujarat by the Indian Institute of Management, Admedabad, revealed that nearly 67% of the owners were of medium socioeconomic status, and only 26% were from the low-income group (Moulik and Srivastava 1975). The individual families who owned gas plants had, on an average, 10 ha of land and 10 head of cattle. In another survey by the Dena Bank in Gujarat, most owners had an annual income of more than US $1100 (many over US $2800), and their primary occupation was agriculture (Dena Bank 1975). They were all literate and nearly 40% had subsidiary occupations such as a business or a service. Others of equivalent social status had no such subsidiary occupation.

Similarly, in the State of Haryana, which has the largest number of biogas plants, five villages were surveyed. Of the 12 biogas plants in these villages, five were in one village — this was attributed to the enterprising character of its inhabitants. Out of a sample of 835 households, 681 had animals; a rough breakdown follows: landless 39% (farm labourer 36%, business 1.5%, services 1.5%); marginal farmers (up to 1 ha) 13.5%; small-scale farmers (1-2 ha) 15.3%; lower medium (2-4 ha) 17.5%; upper medium (4-6 ha) 5.5%; and large (over 6 ha) 9.2%. The biogas plant owners are from the last three categories, which represent only about one-third of the population. This situation is generally repeated in other Asian nations. Essentially, it is the rich who have installed biogas plants. Many factors have made it difficult for the poor to use biogas plants.

Motivation

The motivation for biogas plant installation varies with countries, but it is difficult to avoid the conclusion, at least on Indian experience, that past demand for biogas has been generated by external inducement.

Occasional kerosene scarcities, irregular supplies of petroleum, and scarcity of firewood due to intensive cultivation, as well as the problems of burning firewood during the rainy season, etc., may well have induced some individuals in India to install biogas plants. According to the Dena Bank survey (1975), about 93% of the owners installed their units for cooking. The smokey flame from the traditional fuel (cattle dung) blackens the kitchen and utensils and affects the eyes. According to a survey of 56 gobar gas plants in Uttar Pradesh, biogas use has reduced the eye infections of housewives, saved time, increased the life of utensils, and improved the cleanliness of the house and the dress of the women (Sathianathan 1975).

Although there is an increasing awareness of the high value of the digested slurry as manure, the value of the gas seems to have been the prime attraction in the past. The main advantage as fertilizer is perceived to be that the digested slurry can be used to speed up the process of composting other wastes and thereby increase the volume of compost produced.

When the disposal of dung becomes a problem, as in large urban areas, the biogas plant is seen as coming to the rescue. The digested slurry from the biogas units in Madras and Bombay is fed into the city drainage systems. Some families close to Calcutta built their digesters in response to complaints from their neighbours about the smell of the dung and the number of flies and mosquitoes. Others were attracted by the saving on septic tanks by connecting toilets to the biogas units (one firm in Sangli, Maharashtra, supplies prefabricated toilets along with the biogas units).

The need for gas was important to the installation of biogas at Tulsisham Temple in the Gir Forest, Gujarat. Located in the jungle and having no power source, its 300 head of cattle feed a digester that generates electricity and the power to lift water. The Temple uses wood for cooking because of the need for mass cooking on short notice: this highlights the need for an alternative fuel. If, all of the biogas from a community plant were used for irrigation or small industry, and electricity came to the area, alternate uses for the biogas would have to be found. Some large farms with power are nevertheless interested in biogas for lift irrigation or for greater independence from the rural power systems, in which power cuts are common.

Biogas is also successful in delta areas that have an adequate number of cattle, but no forests to supply firewood. Multiple cropping in these areas creates a demand for fertilizer. The relative success of biogas in the Krishna and Cauvery deltas appears to confirm this.

The educated and well-to-do in Andhra Pradesh were reported to be attracted by the convenience of the gas, but in contrast the lower middle class workers are motivated by the value of the digested slurry as fertilizer. Other factors that are generally not considered by individuals are important to state or national governments and these include environmental and deforestation control, public health advantages, and cost savings through recycling of refuse.

The key motivation in Thailand stems from the desire to use the gas as an alternative to expensive charcoal, although pollution control could be a motivation for piggeries, particularly in southern Thailand. Most units in Indonesia are demonstration plants and wood is plentiful; only overpopulated Java, which is facing a deforestation problem, might be interested in biogas for fuel. In the Philippines, the gas is again the main attraction, but the easy availability of firewood means that pollution control is likely to be the motive in the future. In Korea, gas is used for cooking and to save compost materials like straw and forest products from being burned for fuel. Pollution control is the prime force in the propagation of biodigesters in Japan.

Few people drew attention to the advantages arising from the release of household labour for the other productive activities (as in the case of a Korean farmer who increased silk output). Although apparently insignificant, this may be, or could become, an important motivating factor.

Nonadoption

The survey showed that many reasons quoted for nonadoption were often too simplistic: a complex set of interactions are involved and these vary between areas and countries. Many of the reasons for nonadoption are associated with three main problems.

First, people just do not have sufficient resources (of capital, input materials, land, or time) to run plants efficiently. Lack of cattle, due either to different agricultural systems or the increase of mechanization, is seen as a major constraint to biogas adoption in Thailand, Korea, and among the poor sections of Indian society. This is in spite of the fact that enterprising people have managed to run 2.8-m^3/day plants on one animal and a calf. Inadequate water supply is another input that prevents the spread of plants. Lack of space either for the plant or for slurry pits is similarly often cited as a constraint (Dena Bank 1975). Lack of cash liquidity forms another barrier to the purchase of biogas plants. The absolute size of the capital required, together with problems of cash flow, certainly rule out biogas for the poorer sections of society.

Second, the returns to investment are seen as too low in relation to other uses of the resources. The returns to the individual are often thought to be inadequate because they are notional rather than in cash; they are in the form of savings in the use of other resources rather than in direct sales. The value of using the digested slurry is often only available to the farmer who can use it on his own land, but in India 70% of household milch cows are owned by landless people (National Dairy Research Institute, Karnal, India). Low returns also occur when the gas is not particularly valued because of the availability of other sources of heat, particularly wood; the Indonesian program may be limited in this way. In Korea, where 90% of villages are said to be supplied with cheap electricity (US $2 per family per month), the cost of the alternative biogas, with a capital cost of US $150 and low gas production in the cold winter, may be less preferable.

The spatial arrangements of communities form the third major problem area preventing the adoption of biogas. Returns to biogas often become negative if the digested slurry has to be transported a considerable distance to the fields, or if the gas is produced at some distance from where it is to be used (such as in kitchens); similarly, problems arise if the cattle are moved away from the digester in summer (Dena Bank 1975).

To these three sets of problems must be added the important phenomena such as the limited diffusion of the technical knowledge and experience to run plants (particularly in Indonesia) and the lack of the institutional infrastructure, credit, and extension facilities. In Indonesia, plants were resisted due to the Muslim's attitudes to the use of pig manure. Lack of spare parts and technical problems, however, are not as important as they once were.

Night Soil

In India the psychological and religious barriers to the use of night soil vary considerably. There are instances like in Haryana where 30% of plants have toilets attached to digesters, where some have been discontinued due to pressure from elderly parents. Some households whose plants run on the combined digestion of night soil with cattle dung are unwilling to admit such use. Religious sentiments exist against the use of such gas for cooking food offered during worship either at home or in the temple.

It is difficult to correlate these sentiments with either education or religion: a ladies' college in the South could convince an orthodox family to use night soil by arguing that fire has no unholiness; but the college itself was forced by the students to use the gas only in the chemistry laboratory. Religious feelings cause objections in parts of Uttar Pradesh, West Bengal, and the South, but other religious areas like Gujarat and Maharashtra have shown a great willingness to use night soil.

In Thailand, the Philippines, and Korea, there are likely to be psychological barriers, but BUSTI considered that such resistance is not likely in Indonesia, where human waste is currently used as fish feed. Such psychological inhibitions can disappear in time.

However, night-soil use does warrant certain precautions: worms and parasites are more commonly present in night soil than in cow dung, and sludge brought out undigested can have an offensive smell and cause health problems.

Community Plants

The idea of community plants provides a possible means for bringing the benefits of biogas systems within reach of poorer sections of the rural population. Although no community plants are now operating in India, a number of large-scale units for schools, villages, prisons, and other institutions are being considered. However, these cannot be strictly classified as community plants as a single institution owns the inputs, and controls the outputs.

During the survey there was little positive response to community plants because many felt that Indians were too individualistic. Most cooperative ventures succeeded only as long as there was positive leadership, such as the case of the Khiroda Panchayat (Maharashtra) community plant using night soil for street lighting that failed after the transfer of its most enthusiastic worker. The Harigan Cooperative in Mahishal, Maharashtra, has decided against community biogas in its developments even though such a plant was used initially. Suggestions for the promotion of community plants are receiving considerable attention, and the community plant as a commercial venture is being considered, as is the formation of a biogas corporation at the national or state levels.

Positive aspects of a community plant include large-scale efficiency for rural power generation, industry, water pumping, etc. A number of social and technical problems would have to be solved, however. People would have to pool animal livestock waste resources, use community latrines, avoid excessive use of water, and not add disinfectant to the wastes. Technical problems would involve the equitable distribution of the biogas produced (as well as the slurry). Above and beyond this would be the cost and problems of management.

The collection of input wastes might pose a problem, and as well, cow dung is used for domestic fuel, in brick kilns, in rural house construction, etc. Thus it is essential to study these alternate uses and the seasonal fluctuations of their supply. There is some concern that the demand for biogas might deny its availability to existing users: the poor thus deprived of dung, and unable to use gas, might turn to wood, and so cause deforestation.

Thailand, Indonesia, and the Philippines have no community plants and are faced with problems, such as lack of cooperative spirit, similar to those in India. However, the National Housing Authority of the Philippines is planning large plants for new settlements, and Korea will build eight large digesters in selected villages.

Extension

As mentioned earlier, the extension program in India was greatly stimulated by the energy crisis. The State of Haryana alone set up over 12000 plants in about 2 years and it leads all other states in extension work. Factors contributing to this success include planning and execution at the grass-roots level, an intensive media campaign, a fair price structure, and accessible bank loans. The Haryana State Government considers the influence of successful plants crucial in the creation of new demand; thus, farmers who own digesters are asked to demonstrate them to prospective owners. Similar methods are working in Sangli District in Maharashtra, and in the Punjab.

Another important factor in extension work has been the 'approved supervisor,' who is an artisan trained and authorized by the KVIC to act as a biogas agent. Although primarily a salesman, he is sometimes much more: he helps arrange loans for the construction of the digester, sometimes employing trained village youths for the job; afterward he counsels and advises technically on the day-to-day operation of the digesters, being much more visible and accessible than any government official could be. At present there are over 400 trained supervisors.

The need for local workshop facilities, standardization, and easy availability of spare parts was stressed by many owners during the survey. It is still difficult to get biogas burners and lamps tested in approved government laboratories.

The opportunity to examine extension activities in other Asian countries was severely limited: clearly it is difficult to draw conclusions from this scattered set of observations. The identification of visible 'leaders' to initiate a process of diffusion was quite successful in some cases. However, there was no clear evidence of whether these diffusion processes reached 'downward' toward the poorer strata of society or merely 'sideways' to similarly well-off families. One issue is perhaps clear: to be effective, the extension of knowledge about biogas plants must operate very closely with services providing accessible and usable credit or subsidies, and with technical services providing the necessary equipment and guarantees of maintenance and trouble-free service.

Credit and Subsidy

Credit procedures for biogas plants are complicated by the very low resale value of the plants and the consequent reliance on third party guarantees. Furthermore, advances for biogas units by Indian banks are based on Government cost estimates that are uniform for the entire country. According to several owners these are lower than the actual costs in some areas because of rocky soil or simply higher labour and material costs. Interest rates on these advances are high, and there is much 'running around' involved. Because of this, a number of middle-class owners consider loans a burden and prefer to raise their own funds. There was a phase of interest-free loans early on, but with present rates up to 14%, many owners wanted a reduction in interest rates and extension of the repayment period to 10 years. Some even suggest that interest should only be charged in the case of defaults.

Although family plants can operate with two animals, most banks insist that the borrower have at least five or six animals, with a minimum of 2 ha cultivated agricultural land. It is clear from these conditions that the credit system is meant for the wealthier classes, and it also indirectly reflects who the biogas owners are.

From the banks' point of view the rate of interest is the same as that normally applied to other agricultural advances (4% over bank rate), and most of the borrowers are relatively well-to-do.

The question of the length of the loan repayment period is complicated by the KVIC experience that 95% of the 10-year loans approved have defaulted. Nevertheless many banks advance the loan on personal guarantees without insisting on other securities.

The former outright subsidy of US $34 (Rs 300) was replaced by one equal to 25% of the plant cost — this has now been reduced to 20%. Some marginal farmers and employed plant owners said they were attracted by the subsidy, but the Dena Bank survey in Gujarat reported that subsidy was not a major factor in attracting plant owners. In fact, according to the banks the subsidy should be withdrawn completely because the benefits go to the well-to-do; if continued, it should be confined to the marginal farmers. Curiously, the subsidy is not given for plants totally operated on night soil.

In the view of the Reserve Bank of India (1976), biogas plants of all sizes are profitable, and the continuation of subsidies can be supported if they are considered as an income transfer both from the present generation to the future one (for the conservation of fossil fuels) and as a transfer from urban to rural areas (Sanghi et al. 1976). During the survey it was found that the motivation to invest in biogas was rarely based purely on economic grounds.

Very little information is available on the credits and subsidies in other countries. Thailand has abolished subsidies, and the withdrawal of the subsidy in Korea has drastically affected further installation of digesters. On the other hand, Sri Lanka is planning to establish financial subsidies and in Pakistan the government supplies gasholders free to farmers who build their own digesters.

Other Benefits

During the survey some officials referred to the indirect social benefits resulting from the biogas extension program. These included a spirit of self-reliance, an increased diffusion of metallurgical and technical skills, and a general rise in the standard of living and cleanliness.

Some General Implications

It is clear that biogas systems in the Asian region could provide fuel and fertilizer substitution, waste recycling, pollution control, and improvement of sanitary conditions. What is not so clear is how significant these contributions are now, and will be in the future. Nor is it clear who benefits from the exploitation of the technology. Extreme positions are taken by both the enthusiasts and the sceptics, with both sides tending to disregard the facts.

However, a number of people have attempted to make serious, objective assessments of the social and economic potential of the technology in Asia. Most relate to India and the system most widely used there — the family-sized plant based on cattle dung waste. Little information is available from other countries (see Chapter 2; also Berger 1976; ICAR 1976; Moulik and Srivastava 1975; Prasad et al. 1974; Reserve Bank of India 1976; and Sanghi and Dey 1976), and even the best of these studies ignore or misinterpret some of the social and economic issues discussed here. Even when taken together, these papers provide almost no guidance for judgements and policies in other Asian countries.

The main objective of biogas investment in rural Asia should be to improve the distribution of income by serving the needs of a wide range of social groups. Depending on the local socioeconomic conditions, biogas investment will have its own order of priorities: for example, it may become an attractive opportunity only when certain other investments such as irrigation have been carried out (see Chapter 2). The present ownership pattern reveals that biogas systems can be afforded only by the relatively well-off. Technical as well as socioeconomic considerations should dictate the operation of large community plants.

Problems of Evaluation

A considerable number of potential benefits from biogas systems have been suggested, but the problem remains of evaluating the conditions under which these benefits can be reaped.

Macroevaluations often assume that most of the available inputs will go directly into the digester, but microevaluations suggest that alternative uses of inputs and seasonal fluctuations limit their availability in practice. Local factors like climate, cropping pattern, terrain, and social practices will influence not only technical design, but also costs and benefits. Most of the present evaluations lack reliable data at the microlevel and suffer from an underestimation of the costs and overestimation of the benefits (Taylor 1976). Some evaluations also make the error of 'double counting.' For example, when organic wastes are added to the slurry pit and composted with digested slurry, we cannot apportion the total value of the compost to the biogas plant because had the organic wastes been composted, they would have given an almost equal quantity of manure. The biogas plant speeded up the composting process: the value to a rural economy of this speeding up of the process is a complex question.

Present evaluations of biogas in India assign greater value to the slurry than to the gas (ICAR 1976; Moulik and Srivastava

1975), and this is so even without considering the organic humus benefits. The ICAR study presumes that the value of the manure is 2.3 times the value of the gas. Even Disney (1976) does not give the sludge any value other than as nitrogen even though his study is on the economics of fertilizer production. The best way to assess the true value of the slurry would be to measure the extra output of crops, algae, or fish (e.g. slurry is 13% more effective than farmyard manure — Berger 1976; Idnani and Varadarajan 1974).

Ambiguities in the data abound; if we want to find out what the fuel value of biogas is, we must measure methane content, but many assessments of increased gas production ignore this fact. Dung output will vary widely with the breed of animal, its food etc., and this is why generalizations on the number of animals to support a given size of plant are often misleading. There can be no generalizations about the price of inputs as they will vary with season and location. Even family wealth cannot be defined in terms of land ownership as the land may be infertile or their actual standard of living may continue to be low.

So far only a small number of known designs have been built and tested. There remains great scope for improvements and cost reduction, yet even the existing system is reported to be highly cost-efficient. A good cost analysis must find not only the different alternatives for fuel and fertilizer, but also pose the question: Is biogas the best use of the available resources? Social and environmental benefits, the depletion of nonrenewable resources, and fluctuations occurring outside the system have rarely been taken into consideration (an energy crisis can considerably alter the benefits of biogas).

The possible role of Government subsidy must be viewed from the overall context of socioeconomic development and self-reliance. A baseline may be needed, and a scenario could describe what would happen in a particular village under various assumptions if electrification took place, or if biogas plants were installed, a sugar mill built, or any combination of such factors (Taylor 1976).

Summary

Asian biogas systems are characterized by great diversity even though only a limited number have actually been built. Most are used for family cooking, although other uses are on the increase. But even so, burner and appliance efficiency has still received inadequate attention.

The greatest benefits from biogas systems are to be derived from the manurial value of the slurry; however, this fact is not well known outside India and China. Even in India, the ability of biogas digesters to convert part of the organic nitrogen of the feed to ammoniacal nitrogen has not been exploited. The benefits of organic humus and nitrogen 'carry-over' effects of the sludge have still to be investigated and no reliable data yet exist on the increase in crop yields that biogas slurry can produce.

Design and operational improvements must be conducted and the optimum use of the outputs determined. The digestion of cellulosic materials (especially agricultural waste), with the resultant acceleration of the digestion process and reduction of capital costs, would gain a wider acceptance for biogas, particularly in regions that do not have cattle.

The public health control aspects of anaerobic digestion compare with any other feasible night-soil handling techniques. The hardier parasitic eggs are best controlled by physical separation. There is considerable scope for industrial and urban waste treatment as well as for the recycling of livestock waste through the use of biogas.

Motivation for biogas installation varies — governments seem to be most interested in biogas applications for environmental control, foreign exchange savings, and control of deforestation.

Family-size units are owned for the most part, by the well-to-do, as a host of reasons have made it difficult for poorer people to have biogas plants. Nonadoption can be due to psychological and practical problems associated with the handling of various wastes and slurry, or simply a lack of necessary resources (i.e. capital, input materials, land, time). Some prefer to invest elsewhere. Biogas systems can succeed in areas where inputs have low opportunity costs, the alternatives have high opportunity costs, and where plants can be operated with adequate efficiency.

To be effective, an extension program must operate very closely with systems of credits, subsidies, and technical services. Subsidies can be viewed as a transfer payment from the urban rich to the rural poor, or as a transfer payment to the future generations for the conservation of fossil fuels.

The main objective of biogas investment in most parts of rural Asia should be the distribution of income and needs to a wide range of social groups. Large community plants are most likely to achieve this, and their greater efficiency would allow treatment of various types of wastes. Power generation, pumping water, and running rural industry are conceivable uses for biogas in a village — even the waste heat could be effectively used. Such a unit could be run as a commercial venture, but the operation of such community plants is plagued with many social and technical problems.

The criterion of attractive returns on investment matters very little if the necessary capital and means are not available to the villager. Further, the evaluations do not take into account the social and other latent costs of the depletion of nonrenewable resources. An analysis has to consider not merely the different alternatives for meeting fuel, fertilizer, and other needs but also whether investments in biogas are the best use of available resources. The economics of biogas systems is highly location-specific and it is essential to identify rural zones with the right potentials and socioeconomic environment to maximize the returns to the individual, the rural community, and the nation as a whole.

The timely funding from both the International Development Research Centre and the Indian Council of Social Science Research, and the consent and encouragement of Dr B.K. Madan, Chairman, Management Development Institute, enabled me to undertake this study. My thanks are due to all of them. I am particularly grateful to Dr Alicbusan, Dr (Mrs) Revades Deemark, Dr W.D. Han, Dr Ashok Jain, Mr H.R. Sreenivasan, and Dr G.P. Sudirjo, who took great pains in arranging my survey visits, and to the many people in different countries who freely shared their experience and educated me on the subject. This work greatly benefited from the advice and encouragement of Mr R. Martin Bell and Dr C.H.G. Oldham.

Appendix 1

Continuous digestion: typical gas yields

Component	Experimental conditions	Total gas production		Methane/ carbonydrate (%)	Properties of component in feed (%)	Cellulose destroyed (%)
		ft^3/lb added	ft^3/lb dest- royed			
Green garbage (76.5% garbage + sludge)	T = 37 °C, stirred θ = 30 days LR = 0.077 lb/ft^3/ day nonacclimatized		17 2.4 l/ day	58/42		74.3% VS reduction[1]
Green garbage (100%)	T = 37°C, stirred θ = 30 days LR = 0.154 lb		14	60/40		65% VS destroyed[1]
Paper pulp (50% sludge 50% pulp)	T = 37 °C θ = 30 days		12.9			90.3% cellulose destroyed 64.7% total solids destroyed[1]
Green garbage	T = 37 °C θ = 30 days 3L digester LR = 0.77 lb VS/ft^3/day	8.8	13.9	54.7/45.3	100	92.4[2]
Kraft paper		9.2	12.1	66.5/33.5	60	92.4
Newspaper		7.5	13.0	69.5/30.5	30	41.7
Garden debris		7.8	11.9	69.5/30.5	50	78.5
Wood		4.3	7.7	69.7/30.3	60	
Chicken manure		5.0	17.1	59.8/40.2	100	
Steer manure		1.4	8.7	65.2/34.8	100	
Sewage sludge		9.7	15.2	64.5/35.5	0	

[1]SERL Report No. 67 (1967).
[2]Klein (1972).

Appendix 2

Continuous/batch digester rates: some models and results

Substrate	Model	Parameters	Temperature effect	Reference
Protein (S_0 = 200 mg/l)	1st order	0.023 day^{-1} 0.004 day^{-1} 0.0077 day^{-1}	T = 10 °C E = 72 T = 20 °C T = 30 °C E = 11.8	Ryabov (1974)
Stearic acid	Monod	k_S = 417 mg/l $\hat{q}$ = 0.77 day^{-1}	T = 37 °C	
Palmitic acid	Monod	k_S = 143 mg/l $\hat{q}$ = 1 day^{-1}	T = 37 °C	
Myristic acid (S_0 = 1375 mg/l)	Monod	k_S = 105 mg/l $\hat{q}$ = 0.95 day^{-1}	T = 37 °C	Novak & Carlson (1970)
Oleic acid (S_0 = 1835 mg/l)	Monod	k_S = 3180 mg/l $\hat{q}$ = 4.0 day^{-1}	T = 37 °C	
Linoleic (S_0 = 1835 mg/l)	Monod	k_S = 1816 mg/l $\hat{q}$ = 5.0 day^{-1}	T = 37 °C	
Acetic acid (S_0 = 1568 mg/l θ = 1.5-12 day)	Monod	k_S = 154, 333, 869 mg/l $\hat{q}$ = 8.7, 4.8, 4.7 day^{-1}	T = 35, 30, 25 °C	Lawrence (1969)
Cellulose (S_0 = 13744 mg/l θ = 6-30 days	Monod	k_S = 7530 mg/l COD $\hat{q}$ = 5.4 day^{-1}	T = 37 °C	Chan (1970)
Domestic waste digested with sewage sludge (35-60 °C θ = 4-20 days 3.33 g/l day feed)	1st order[1]	$k_{initial}$ k_{final}/day[2] 0.055 0.003 0.084 0.043 0.052 0.007 0.117 0.03 0.623 0.042 0.99 0.04	T 35 °C 40 °C 45 °C 50 °C 55 °C 60 °C	Pfeffer (1974)
Elephant grass (batch digester, nonstirred, nonacclimatized)	1st order	k = 0.06 day^{-1} at = 0.0526 at	T = 32 °C T = 22 °C	Boshoff (1966)
Cow manure (liquid) (S_0 = 3800 - 15300 mg/l = 10 days)	($Y=G(1-c^{-kt})$) 1st order	k = 0.125 day^{-1}	T = 35 °C	Gaddy et al. 1974
Cellulose extracted from river bed (S_0 = 20000 mg/l) (Batch digester)	Zero order ($R=k$)	k = 0.12 at = 0.23 at = 0.311 at (k = mm gas pressure/h/50 ml sludge fed)	T = 10 °C T = 15 °C T = 20 °C	Springer

[1]Defined with reference to potential gas production
[2]Depends on residence time. Broadly (a) $\theta < 10$ days; (b) $\theta > 10$ days.

Cellulose digestion

System and culture	Initial concentration of cellulose	Cellulose material	pH	Digesting rate	Remarks
Batch, mixed two pure cultures isolated from sewage digester 38 °C, mesophilic	2000 mg/l	Whatman's No.1 filter paper	6.8	(1)260 mg/l/day (2)660 mg/l/day	Two exp. were made with different strains
Batch, mixed culture from sewage digester 25 °C, mesophilic	3120 mg/l	Cellulose in sewage sludge	7.4	142 mg/l/day	pH was maintained with lime
Batch, pure culture isolated from soil and manure, 55 °C, thermophilic	(1) 744 mg/l (2) 2980 mg/l	Absorbent cotton	— —	(1) 149 mg/l/day (2) 426 mg/l/day	Two experiments were made with different strains
Batch, mixed culture from fibrous river sludge, 25 °C	Approx. 20000 mg/l	Cellulose in fibrous river sludge	5.9-6.5	48 to 216 mg/l/day	Range of rates of samples from monthly monitorings
Batch, mixed culture from rumen fluid 40°C, thermophilic	41200 mg/l	Whatman's 6.5, No. 2 filter paper	—	11400 mg/l/day	pH was maintained with $NaHCO_3$
Continuous, mixed culture, sewage sludge	13744 mg/l	Particulate kraft pulp - milled	—	412-1250 mg/l/day	Enrichment culture used θ = 30 - 6 days

Appendix 3

Suggestions for studies in core technology

Plant-scale studies

1. collection of data and evaluation of existing designs, including reasons for failure
2. studies/trials and evaluations (technical and economic) of design modifications (e.g. gas holder design, materials of construction (with corrosion, reliability, cost), mixing, and heating/preheating/insulation)
3. batch-scale plant: study of efficiency, feasibility
4. EVOP studies on selected plants
5. other topics should come forward following laboratory/pilot-scale work as set out below

Basic factors relating to process operation and efficiency

1. rates, yields, and limiting steps as a function of feed material and preparation
2. collect data on effect of operating conditions and constraints on performance

Design modifications etc. (laboratory/pilot-scales) (related where possible to studies above)

1. multistage designs and high loading systems
2. effects of mixing, L/D ratios, etc. on performance
3. plug flow design
4. studies of process constraints, stability, sensitivity

Basic chemistry, microbiology, bacteriology

1. studies on population dynamics, ecology (relate to 'starters', improvements), modifications thereof
2. rate limitations due to microbiological effects
3. shocks, inhibitions with local pollutants
4. pathogen kill studies

Instrumentation

1. development of cheap/robust/appropriate instruments for metering inputs, outputs; temperature; pH; gas metering
2. simple control schemes/strategies

Gas treatment/handling/use (many problems are problems of using known technology, establishing good practice, etc.)

1. purification — design and testing of convenient methods
2. evaluation of potential by-product utilization
3. piping materials (standards, etc.)
4. design, modification of carburetors, etc. for engines
5. burners and lighting

Liquid and solid waste disposal/treatment

Treatment methods with reference to pollution control, nutrient utilization, e.g.

1. heat treatment (pathogens, see earlier)
2. algal lagoons — viability, algal colonies, stability, performance, recovery
3. algal lagoons — kelps, hyacinths
4. fish culture — choice of fish, trials
5. use of CO_2 to promote algal growth

(2, 3, and 4 must include detailed studies of nutrient and toxins cycles)

Systems studies

Studies on systems viability, optimization, constraints around the complete cycle, and alternatives

Bibliography

Acharya, C.N. 1956. With this change, the cow dung plant works better. Indian Farming, 9(6), 10-13.

1958. Preparation of fuel, gas and manure by anaerobic fermentation of organic materials. ICAR Research Series No. 15.

Alich, J.A. and Inman, R.E. 1975. Utilisation of plant biomass as an energy feedstock. In Jewell, W.J., ed., Energy, agriculture and waste management. Michigan, Ann Arbor Science Publishers Inc.

Alicbusan, R.V. 1976. Recent developments in Philippine production and utilisation of biogas. Paper given at NRCP-NIST Symp. on World Development of Biogas Technology, Manila.

1976. The possible improvement of biogas production throught proper utilisation of micro-organisms. In Philippine Council for Agriculture and Resources Research.

Ames, G.C.W. 1976. Can organic manures improve crop production in Southern India? Compost Science, 17(2), 7-11.

Andrews, J.F., Cote, R.D., and Pearson, E.A. 1964. SERL Report 64-11, University of California, Berkeley.

Anon. Biological treatment of industrial waste. Fermentation Research Institute, Japan, undated mimeo.

1974. Fuel gas from agriculture wastes. Trends in Technology, XIII(3), 3, Econ. Dev. Found., Philippines.

1976. The economies of cow-dung gas plants. ICAR, New Delhi.

Anthony, W.B. 1969. Animal waste management. Cornell University Conference, Cornell, Ithaca, N.Y., p. 105

1976, Gobar gas. Appropriate Tech. Development Organization, Pakistan, mimeo.

Atkinson, B. 1975. Biochemical reactors. London, Pion Ltd.

Augenstein, D.C., Wise, D.L., and Cooney, C.L. 1976. Packed bed digestion of solid wastes. Resource Recovery and Conservation, 2, 257.

Bell, Boulter, Dunlop, and Keiller. Methane — fuel of the future. Dorchester, U.K., Prism Press.

Berger, L. 1976. Economics of gobar gas. Report prepared for the United Mission Economic Development Agency, United Mission to Nepal, Butwal, Nepal.

Bertucci, J., Leu-Hing, C., Zenz, D. and Sedita, S.J. Destruction of enteric viruses during anaerobic digestion.

Bhalla, A.S., ed. 1975. Technology and employment in industry. ILO, Geneva.

Boersma, L., Barlow, W.W.R., Miner, J.R., Phinuey, H.K., and Oldfield, J.E. 1975. Protein production rates by algae using swine manure as a substrate. In Jewell, W.J., ed., Energy, agriculture and waste management. Michigan, Ann Arbor Science Publishers Inc., p. 475.

Borchardt, J.A. 1971. Anaerobic phase separation by dialysis technique. In Gould, R.F., ed., Anaerobic biological treatment processes. A.C.S. Series 105, p. 108.

Boshoff, W.H. 1963. The application of methane installation in the tropics. World Crops, 433-434.

1965. Methane gas production by batch and continuous fermentation methods. Tropical Science, V(3), 155-165.

1967. Reaction velocity constants for batch methane formation on farms, notably in the tropics. J. Agric. Sci. 68, 347-349.

Brandt, D., Hontz, L. and Mandels, M. 1973. Engineering aspects of the enzymatic conversion of waste celluose to glucose. AICE Symp. Ser. 133(69), 127.

Briones, A.M. and Briones, A.A. 1976. Organic fertilization — a means to nutrient and waste recycling, In Philippine Council for Agriculture and Resources Research.

Buswell, A.M. and Mueller, H.F. 1976. Mechanism of methane formation. Ind. Eng. Chem. 44(3), 550.

Carr, M. ed., 1976. Economically appropriate technologies for developing countries. London, Intermediate Technology Publications Ltd.

Chan, G. 1973. Integrated rural development. In Workshop on waste recycling systems. Papua New Guinea Dept. of Agriculture, Stock and

Fisheries, Port Moresby, Univ. of Papua New Guinea.
Chung, Po. 1973. Production of methane gas from manure. International Biomass Energy Conference, Winnipeg, Canada
Chung, Po, Wang, H.H., Chen, S.K., Hung, C.M. and Chang, C.I. 1974. Small methane generator for waste disposal. Proc. Third Symp. on Livestock Wastes, U.S. Soc. Agric. Eng., p. 238-240.
Coe, W.B., and Turk, M. Processing animal waste by anaerobic fermentation. Hamilton Standard, Division of United Aircrafts Corp, USA.
Cooney, J. and Wise. 1975. Thermophilic anaerobic digestion. Biotech. Bioeng. 17, 1119.
Cooper, C.M. 1973. Choice of techniques as a problem of technical change. International Social Science Journal (UNESCO), XXV, 293-304.
Dague, R.R. 1968. Application of digestion theory to digestor control. J. Water Poll. Control Fed. 40, 2021-2032.
D.C.S. Nepal. Development and consulting services. Butwal, Nepal, mimeo.
Deemark, R. 1975. Status of biogas production program in Thailand. Paper presented at ESCAP Biogas Workshop, New Delhi.
Dena Bank. 1975. The gobar gas plant scheme, preliminary appraisal (an internal noting). Bombay, India.
Dhua, S.P. 1975. Need for an organo-mineral fertilizer in tropical agriculture, Ecologist, 5(5).
Disney, R. 1976. The economics of gobas gas versus fertilizers: critique of intermediate technology. Dep. of Economics, Univ. of Reading, June 1976, mimeo. (see also Development and Change. 1977, 8, 77-102).
Dugan, G.L.K. and Oswald, W.J. 1968. Mechanisms of anaerobic waste treatment. Proc. Symp. Fed. W.P.C. Admin.
Dunlap, C.E. 1975. A note on the value of cellulose. In Wilke, ed., Cellulose as a chemical and energy resource. Biotech. Bioeng. Symp. Ses. No. 5, p. 73.
ESCAP 1975. Biogas technology and utilisation. United Nations Economic and Social Commission for Asia and the Pacific.
1976. Biogas News Letter, August 1976. United Nations Economic and Social Commission for Asia and the Pacific.
Eusebio Jose, A. 1976. Chlorella-manure, a gas-fish pond recycling system in integrated farming. In Philippine Council for Agriculture and Resources Research.
Eusuf, M. 1975. Prospect for biogas industry in Bangladesh. Paper presented at ESCAP Biogas Workshop, New Delhi.
Fair, G.M., Geyer, J, and Okum, N. 1966. Sewage treatment. New York, John Wiley and Sons.
Fang Chen. 1976. Energy resources: China self sufficient. Economic Times, New Delhi, 15 July.
Farvar, M.T. and Bajrachanya, D.P. 1975. Prospects and problems of bio-gas as an appropriate technology for development in the rural areas — a critical appraisal. ESCAP Workshop on Bio-Gas Technology and Utilization, Manilla.
Feachen, R., McGarry, M., and Mara, D., ed. 1977. Water, wastes, and health in hot climates. London, John Wiley and Sons.
Fermentation Research Institute. Biological treatment of industrial wastes. Fermentation Research Institute, Agency of Industrial Science and Technology, Japan.
Finlay, J.H. 1976. Operation and maintenance of gobas gas plants. Report prepared for the United Mission Economic Development Agency. United Mission to Nepal, Butwal, Nepal.
Finney, J. and Evans 1975. Anaerobic Digestion: the rate limiting process. Science, 190, 4219, 1088.
Fry, J.L. 1973. Methane digesters for fuel gas and fertilizer. New Alchemy Institute, U.S.A.
1975. Methane power plants. D.A. Knox.
Gaddy, J.L., Park, E.L., and Rapp, E.B. 1974. Kinetics and economics of anaerobic digestion of animal waste. Water Air and Soil Pollution, 3, 116-119.
Ghosh, J. and Pohland, F.G. 1974. Kinetics of substrate assimilation and product formation in anaerobic digestion. J. Water Poll. Cont. Fed. 48, 748.
Gotaas, W. 1956. Composting. WHO Monograph, WHO, Geneva.
Government of India. 1965. Report of the energy survey of India committee. Indian Documentation Service, New Delhi.
1974. Report of the fuel policy committee. Indian Documentation Service, New Delhi.
1976. The economics of cow dung gas plants. Indian Council of Agricultural Research, New Delhi.
Government of Pakistan. 1969. Survey report on pattern of use of organic/inorganic manures on various crops in Pakistan. Planning Unit, Agriculture Wing, Ministry of Agriculture and Work, Food and Agriculture Division, Islamabad.

Graef, S.P. and Andrews, J.F. 1973. Mathematical modelling and control of anaerobic digestion, AICE Symp. Ser. 136(70), 101.

Griffin, K. 1972. The green revolution: an economic analysis. U.N. Research Institute for Social Development, Report No. 726.

Griffin, S. 1975. A communication approach to determining biogas needs and application at the farm level. Report prepared for ESCAP Workshop on Biogas Technology and Utilization, Manilla.

Hadjitofi, M. 1976. The kinetics of gas production in the anaerobic digestion of cellulose wastes. PhD Thesis, Univ. of London.

Hart, S.A. 1963. Digestion tests of livestock wastes. J. Water Poll. Cont. Fed. 35(6).

Hattingh, W.H.J. and Toerien, D.F. 1969. Anerobic digestion: I. the microbiology of anaerobic digestion. Water Res. 3, 385.

Heukelekian, H. and Mueller, P. 1958. Sewage and Industrial Wastes. 30, 1108.

Hitachi Plant Engineering and Construction. A report on trials with urban wastes (Japanese text). Hitachi Plant Engineering and Construction Co. Ltd., Japan.

Hobson, P.N. 1973. The bacteriology of anaerobic sewage digestion. Process Biochemistry, 19-25.

Hobson, P.N., Bousfield, S., and Summers, R. 1974. Anaerobic digestion of organic matter. Crit. Rev. in Enviromental Control, 131-191.

Hobson, P.N., Robertson, A.M., and Mills, P.J. 1975. Anaerobic digestion of agricultural wastes, A.R.C. Research Review, 1(3), 82-85.

ICAR 1976. The economics of cow dung gas plants. Indian Council of Agricultural Research, (ICAR), New Delhi.

Idnani, M.A. and Varadarajan, S. 1974. Fuel gas and manure by anaerobic fermentation of organic materials. ICAR Technical Bulletin No. 46.

Idnani, M.A. and Acharya, C.N. 1963. Biogas plants, their installation, operation, maintenance and use. ICAR, New Delhi.

Ifeadi, C.N. and Brown, J.B. 1975. Technologies suitable for recovery of energy from livestock manure. In Jewell, W.J., ed., Energy, agriculture and waste management. Michigan, Ann Arbor Science Publishers Inc.

International Council of Scientific Unions. 1975. Scientific Committee on Problems of the Environment, Environmental Impact Assessment, Munn, R.E., ed., Toronto, Canada.

Institute of Agricultural Engineering and Utilization 1976. Agricultural Engineering (ORD), Republic of Korea. Division of Design. Material for methane gas producing facilities.

Irvin, W. 1976. Modern cost-benefit methods: an introductory textbook. Institute of Social Studies, The Hague.

Islam, M.N. 1976. Production of biogas from diffeıent types of wastes. Paper presented at the Seminar on Food and Energy from Unconventional Sources with Application of Nuclear Techniques, Dacca.

Jewell, J., ed., 1975. Energy, agriculture and waste management. Ann Arbor Science Publishers Inc., Michigan, U.S.A.

de Jounge, 1976. The Toa-throne — a new compost toilet. Compost Science, 17, 16.

Khadi Village Industries Commission. 1975. Gobar gas: why and how. Khadi Village Industries Commission, Directorate of Gobar Gas Scheme, Bombay.

Kamata, S. and Uchida, K. Studies on livestock excreta disposal by methane fermentation. (II) Optimum load and comparison between mesophilic and thermophilic fermentation. Dept. of Animal Hygiene, Nippon Veterinary and Zootechnical College.

Klausher, S.D., Zwerman, P.J., and Scott, T.W. 1971. Land disposal of manure in relation to water quality. Cornell University, Agronomy Dept. Paper No. 926. Cornell Univ. Conference on Agric. Waste Management, p. 36-44.

Klein, S.A. 1972. Anaerobic digestion of solid wastes. Compost Science, 6, 11.

Klopfenstein, T. and Koers, W. 1973. In Inglett, G.C., ed., Symp. on Processing Agric. and Municipal Wastes, Westport, Conn.

Lauer, D.A. 1975. Limitations of animal waste replacement for inorganic fertilizers. In Jewell, W.J., ed., Energy, agriculture and waste management. Michigan, Ann Arbor Science Publishers Inc., p. 409.

Laura, R.D. and Idnani, M.A. 1971. Increased production of biogas from cow dung by adding other agricultural waste materials. J. Sci. Fd. Agr. 22, 167-171.

Lawrence, A.W., and McCarty, P.L. 1976. Kinetics of methane formation in anaerobic treatment. J. Water Poll. Control Fed., R1.

Lawson, G.H. and Windle, D.W. 1974. Tables for discounted cash flow. Longman, London.

Lee and Kim. 1975. Studies on the biogas production and its storage. College of Agriculture, Suweon.

Leui, S.S. 1975. Production of methane from farm wastes. Paper presented in ESCAP Biogas Workshop, Manila.

Lessage, E. and Abiet, P. 1952. Gaz de fumier. La

Diffussion Nouvelle du Livre, Soisons, France.
Little, I.M.D. and Mirrlees, J.A. 1974. Project appraisal and planning for developing countries. Heinemann Educational Books, London.
Loehr, R.C. 1971. Alternatives for the treatment and disposal of animal wastes. J. Water Poll. Control Fed. 43, 668.
Makhijani, A. 1976. Energy policy for the rural Third World. Int. Inst. for Environment and Development.
Makhijani, A. and Poole, A. 1975. Energy and agriculture in the Third World. Ballinger Publication Corporation, Cambridge, Mass.
Maki, L.R. 1954. Antonie van Leeuwenhoek, 20, 185.
Malina, J.F. and Miholits, E.M. 1967. New developments in the anaerobic digestion of sludges. J. Water Poll. Control. Fed. 40, 42.
Malynizc, H. 1973. In Workshop on Waste Recycling Systems, Papua New Guinea, Dept. of Agriculture, Stock and Fisheries, Univ. of Papua, New Guinea, Port Moresby.
Mann, I. 1962. Processing and utilization of animal by-products. FAO, Rome.
Mardon, C.J. 1976. An assessment of small-scale methane — algal systems in the Asian Pacific area. Report prepared for Assoc. for Scientific Corportation in Asia, Melbourne.
Marstrand, P.K. et al. 1974. Social and ecological evaluation of industrial development. Science Policy Research Unit, Univ. of Sussex, mimeo.
McBee, R.F. 1948. Journal of Bacteriology, 56, 653.
McCann, D.J. and Saddler, H.D.W. 1976. Photobiological energy conversion in Australia. Search, 7(12), 17-23.
McCabe J. and Eckenfelder, W.W., eds. 1958. Biological treatment of sewage and industrial wastes, Vol. 2. New York.
McCarty, P.L. 1964. Anaerobic waste treatment fundamentals, II, III. October, November, Pub. Works.
McCarty, P.L. 1971. In Faust, S.J. and Hunter J.V., eds., Organic compounds in aquatic environments. Chap. 21, p. 495.
McGarry, M. 1971. Water and protein reclamation by sewage. Process Biochem. 6(1), 50-53.
1976. Health aspects and the potential for transmission of human diseases through the biogas plant. IDRC, Canada.
McGarry, M., and Stainforth, J. 1978. Compost, fertilizer, and biogas production from human and farm wastes in the People's Republic of China. IDRC, Ottawa, IDRC-TS8e.
McGarry, M., Lin, C.D., and Merto, J.L. 1972. Photosynthetic yields and by-product recovery from sewage oxidation ponds. 6th Conf. Int. Ass. Water Poll. Res., Jerusalem, p. 1, Pergamon Press.
Meynell, P.J. 1976. Methane, planning a digester. U.K., Prism Press.
Mohan Rao, C.J. 1974. Scientific aspects of cow dung digestion. Khadigramodyog (India).
Mosey, F.E. 1974. Anaerobic biological treatment. Inst. of Water Poll. Control, Symp. on Treatment of Wastes from Food and Drink Industry, Newcastle-Upon-Tyne.
Moulik, T.K. and Srivastava, U.K. 1975. Biogas plants at the village level: problems and prospects in Gujarat. Indian Institute of Management, Ahmedabad.
National Academy of Sciences. 1977. Making aquatic weeds useful: some perspectives for developing countries. N.A.S., Washington, D.C.
NASA. 1974. NASA, Technical Memo No. 72725.
Neyeloff, S. and Gunkel, W.W. 1975. Methane-carbon dioxide mixtures in an internal combustion engine. In Jewell, W.J., ed., Energy, agriculture and waste management. Michigan, Ann Arbor Science Publishers Inc.
Novak, J.T. and Carlson D.A. 1970. J. Water Poll. Control Fed. 42, 1932.
Organization for Economic Co-operation and Development. 1968. Supply and demand prospects for fertilizers in developing countries. OECD, Development Centre Studies, Paris.
ORD. 1976. Utilization effects of methane gas as a rural fuel source. Suweon, Republic of Korea.
Obias, E. 1975. Waste utilization at Maya Farms. ESCAP, Manila.
1976. Biogas system at the Maya farms. In. Philippines Council for Agriculture and Resources Research.
Pantastico, J.B. 1976. Chlorella production and utilization in the Philippines. In Philippines Council for Agriculture and Resources Research.
1976. Historical account of plant nutrition and soil fertility studies. In Philippines Council for Agriculture and Resources Research.
Patel, J.J. 1975. Gobar gas plant, its development, present status, and future. Presented at ESCAP Workshop, New Delhi.
Pathak, P.V. and Colah, J.J. 1976. Methane from biogas. I.I.T. Bombay, Science Today (India)

People's Publishing House. 1974. Experience on health management of dung in rural areas — a collection of materials, Peking.
Perrigo, L.D. and Dermmit, T.F. The implications of new food, energy and waste utilization concept. Battelle Memorial Institute, Pacific North West Labs, Richland, Washington.
Pescod, M.B. 1971. Sludge handling and disdisposal in tropical developing countries. Asian Institute of Technology, Bangkok, Thailand.
Pfeffer, J.T. 1968. Increased loadings on digesters with digested solids. J. Water Poll. Control Fed. 40, 1020.
1974. Temperature effects on anaerobic fermentation of domestic refuse. Biotech. Bioeng. XVI, 771-787.
Pfeffer, J.T., Leiter, M., and Worlund, J.R. 1967. J. Water Poll. Control Fed. 39, 1305.
Philippines de la Salla University. Economic implications of the use of biogas as a source of energy in the Philippines. A Preliminary Study, mimeo.
Pimentel, D. 1975. World food, energy, man and environment. In Jewell, W.J., ed., Energy, agriculture and waste management. Michigan, Ann Arbor Science Publishers Inc.
Pohland, F.G. and Ghosh, S. 1971. Biotech. Bioeng. Symposium, 2, 85.
Prasad, C.R., Prasad, K.K., and Reddy, A.K.N. 1974. Biogas plants: prospects, problems and tasks. Economic and Political Weekly, IX (32-34), Special Number.
Production Team of T'ang Ngan. Experience and management of human excreta by Province of Kwangsi. Production Team of T'ang Ngan, District of Fu Sui.
Pyle, D.L. 1977. Pyrolysis of wood: a working paper. Intermediate Technology Development Group, London.
Pyle, D.L. and Freeman, C. 1977. Methane production by anaerobic fermentation: an annotated biography. Intermediate Technology Development Group, London.
Republic of Korea, Office of Rural Development, Oct. 1973. Effects of methane gas utilization as a rural fuel. Korea, mimeo.
Republic of Korea, Office of Rural Development, July 1975. Methane gas utilization as a rural fuel in Korea. Korea, mimeo.
Research Office for Parasitic Disease Prevention. 1973. Experimental research on excreta disposal by biogas plants. Preliminary Report, Research Office for Parasitic Disease Prevention, Province of Szechuan.
Reserve Bank of India, 1976. Financing of gobar gas plants by banks. Reserve Bank of India Bulletin.
Richard, C. 1975. Manuel de gobar reutilisation des effluents d'origine annuale dans la Pacifique Sud. Commission du Pacific Sud, Noumea, Manuel No. 13.
Robb, J. and Evans, P.J. 1976. Recycling of cereal straw. U.K./ISES Joint Conference on Solar Energy in Agriculture, University of Reading, p. 33.
Ryabov, A.K., Nabivanets, B.I., Oleynik, G.N., Smikun, T.Y., and Podgayevskaya, L.V. 1974. Effect of temperature and oxygen concentration on the rate of breakdown of soluble protein in natural water. Hydrobiological J. 10, 8.
Sanghi, A.K. and Doy, D. 1976. A cost benefit analysis of biogas production in rural India: some policy issues. Centre for Biology and Natural Systems, Washington University.
Sathianathan, M.A. 1975. Biogas — achievement and challenges. Association of Voluntary Agencies and Rural Development, New Delhi.
SERL, 1967. Comprehensive studies of solid wastes management. Sanitary Eng. Research Laboratory, Annual Report No. 67-7, Univ. of California, Berkeley.
Shaw, D. 1973. Some factors affecting the growth of algae in waste ponds. Papua New Guinea Workshop, p. 46.
Singh, K.K. 1978. Janta biogas plants: technological administrative feasibility report. PRAI, State Planning Institute, Gobar Gas Research Station, India.
Singh, Ram Bux. 1971. Biogas plant. Gobar Gas Research Station, Ajitmal, India.
1971. Some experiments with biogas. Gobar Gas Research Station, Ajitmal, India.
1973. Biogas plant — Designs with specifications. Gobar Gas Research Station, Ajitmal, India.
Smith, P.H. and Mah, R.A. 1966. Kinetics of acetate metabolism during sludge digestion. Applied Microbiology, 14, 368.
Solly, R.K. and Yarrow, R. 1975. Biogas on Fiji: economic and social factors and future prospects. Paper prepared for the ESCAP Workshop on Biogas Technology and Utilization, Manilla.
Solly, R.K. et al. 1975. Methane digestors in Fiji. Discussion paper for ESCAP Workshop on Biogas Technology and Utilization, New Delhi.
Sonoda, Y., Ono, H. and Nomura, S. 1965. Effects of digested sludge in mesophilic methane fermentation. Fermentation

Research Institute, Japan.
Sonoda, Y. and Seiko, Y. Degradation and toxicity of various compounds in anaerobic digestion. Fermentation Research Institute, Japan.
Springer, A.M. 1972. Investigation of the environmental factors which affect the anaerobic decomposition of fibrous sludge beds on stream bottoms. Institute of Paper Chemistry, PhD Thesis.
Squire, L. and van der Tak, H.G. 1975. The economic analysis of projects. Published for the World Bank by the John Hopkins Univ. Press.
Sri Lanka, Industrial Development Board, 1976. Biogas — why and how?
Stewart, F. 1973. Choice of technique in developing countries. In Cooper, C., ed., Science, technology and development. London, Frank Cass and Co. Ltd., p. 99-122.
Stidham, H.C., Brix, A.T., Dawson, G.N., and Heinemann, W.W. 1973. Cattle manure treatment and refeeding. Pacific N.W. Regional Meeting, Am. Inst. Chem. Eng., Vancouver
Stranks, D.W. 1956. Canadian Journal of Microbiology, 2, 56.
Sudirjo, G.P. and Kismomihardjo, S. 1975. Report on utilization of biogas in Indonesia. Paper presented at ESCAP Biogas Workshop, New Delhi.
Sumicad, R. 1975. ESCAP Meeting on Biogas, Paper 14T/BG/2/9, Manila.
Szekely, M. 1975. Towards a method for the socio-economic evaluation of new technologies for feeding cattle with sugar cane. Progress Report of an IDRC Project, Mexico, mimeo.
Taiganides, E.P. and Hazen, T.E. 1966. Properties of farm animal excreta. Trans. Am. Soc. Agric. Energy, 374-376.
Takatani, T., Takemura, M., Tegima, T., and Osaka, Y. 1975. Studies on fuel gas production from organic wastes with bioconversion in Japan. Paper presented at International Congress of Scientists on Human Environment, Kyoto, Japan.
Tatom, J.W., Chiang, T.I., Harahap, F., Mischidim, A.R., and Wirojosumato, H. 1977. Pyrolytic conversion of agricultural and forestry wastes to alternate energy sources in Indonesia. Feas. Study for USAID, Proj. A-1914, AID/ASIA-C-1203, Feb. 1977.
Taylor, C. 1976. Critical factors in economic evaluation of small decentralized energy factors. South Asia Dept. of World Bank, Washington.
Tietjen, C. 1975. From biodung to biogas — historical review of European experience. In Jewell, W.J., ed., Energy, agriculture and waste management. Michigan Ann Arbor Science Publishers Inc. p. 247.
Torpey, J. and Melbinger, A. 1967. J. Water Poll. Control Fed. 39, 1464.
Trevelyan, W.E. 1975. The methane fermentation: a discussion paper. Tropical Science, 17(4), 193-209.
1975. Renewable fuels: ethanol produced by fermentation. Tropical Science, 17(1).
Tyler, J. 1973. Socio-economic aspects. Summary report of discussions on biogas conducted furing the Workshop on Waste Recycling Systems held at the Univ. of Papua New Guinea, Port Moresby, Papua New Guinea.
UNEP. 1976. UNEP Expert Group Meeting on a methodology for selection of environmentally sound and appropriate technologies. August—September, Nairobi, Kenya.
UNIDO. 1972. Guidelines for project evaluation. United Nations Industrial Development Organization.
UNITAR. 1976. Microbial energy conversion. Conference Proceedings. Schlegel, H.G. and Barnea, J. ed., Pergamon Press.
van de Ryan, S. 1976. The Farallones composting privy. Compost Science, 17, 15.
Ward., G.M. and Seckler, D. 1975. Protein and energy conservation of poultry and fractionated animal waste. In Jewell, W.J., ed., Energy, agriculture and waste management. Michigan, Ann Arbor Science Publishers Inc.
Wolverton, B.C. 1976. Don't waste waterweeds. New Scientist, 318.
1976. Weed holds promise for pollution cleanup. Chemical and Engineering News, 23.
Wolverton, B.C., McDonald, R.C., and Gordon, J. 1975. Bioconversion of water hyacinths into methane gas, Part 1. NASA. TM.X.72725.
Wong-Chong, G.M. 1975. Dry anaerobic digestion. In Jewell, W.J., ed., Energy, agriculture, and waste management. Michigan, Ann Arbor Science Publishers Inc. p. 361.
Worgan, J.T. 1973. Protein production by microorganisms from carbohydrate substances. In Jones. J.G.W., Biological efficiency of protein production, Cambridge University Press.
Yagi, M. 1975. Treatment of livestock waste for pollution abatement in Japan. Paper presented at the Third International Symposuim on Livestock Wastes, University of Illinois.